AF782909

WITHDRAWN

ENGINEERING ASPECTS OF ASBESTOS DUST CONTROL

by

GYAN S. RAJHANS
Chief
Occupational Health Engineering Service
Ontario Ministry of Labour
Toronto, Ontario, Canada

GORDON M. BRAGG
Associate Professor
Department of Mechanical Engineering
University of Waterloo
Waterloo, Ontario, Canada

ANN ARBOR SCIENCE
PUBLISHERS INC
P.O. BOX 1425 • ANN ARBOR, MICH. 48106

TH
7697
.A8
R3

Copyright © 1978 by Ann Arbor Science Publishers, Inc.
230 Collingwood, P. O. Box 1425, Ann Arbor, Michigan 48106

Library of Congress Catalog Card No. 77-92598
ISBN 0-250-40227-0

Manufactured in the United States of America
All Rights Reserved

PREFACE

The dangers of asbestos are well known. That it causes asbestosis, lung cancer, mesothelioma and possibly other diseases is now well documented. At present, an absolutely safe level of dust contamination of the air has not been established, and while this is no different from most other toxic materials, the cancer-causing ability of asbestos means that the lowest possible dust levels must be the aim at all times.

We believe that asbestos dust can be controlled effectively using current technology and that the risks of exposure can be minimized. We have written this book to show how engineers can effectively control exposure levels in the workplace. These control measures can be optimized at minimum cost and maximum safety without significant dislocation of production processes.

The full-time specialist in asbestos dust control will bridle at the thought that his specialty can be placed in a single volume. Of course it cannot. We have confined ourselves, of necessity, to our own experience and research in the field and to published information. We realize that a considerable amount of company confidential information may be available to a designer. We can only encourage him to publish this information for the general good of all asbestos workers.

Asbestos dust control lies within the broad field of industrial hygiene and depends heavily on ventilation. We have attempted to refrain from repeating the basic information in these fields. As a result, the reader will not find in this book any detailed discussion of basic industrial ventilation or other standard control processes. References on these topics are given where necessary. We have also refrained from directing our entire book to a narrow response to the American OSHA and EPA requirements for asbestos dust control, not because most of our experience is Canadian, but also because much of the regulation in this field is evolving and changing, as it should. The documents of the various government bodies are all quite specific in their requirements and have not been repeated here. We have, of course, referred to them at some length where necessary.

We gratefully acknowledge the help and assistance provided by Robert Assad of Université Laval in the early stages of the preparation of this book. Doris Fagan has kindly allowed us to reproduce copyright material from *Asbestos.* George Swallow and Hugh Nelson, among many others, have provided much help and information. The financial assistance of the Science Council of Canada is also much appreciated. To Mary Collins, who gave her help in typing the manuscript, we give special thanks.

G.S. Rajhans
G.M. Bragg

CONTENTS

PART I: ASBESTOS DUST: THE PROBLEM

Gyan S. Rajhans received a BS degree in mining from the Indian School of Mining in 1963 and was awarded the George C. Bateman Fellowship at Queen's University, Kingston, Ontario, to pursue his postgraduate studies in mine ventilation and dust control. After obtaining his MS degree in engineering in 1966, he joined the staff of the Elliot Lake Research Laboratory of the Canadian Department of Energy, Mines and Resources, as a scientific officer. While there, he made studies on various dust sampling instruments, surface areas measurements, and correlation of electronic scanner counts with visual counts of konimeter slides. He also developed a technique to measure radon daughters in uranium mines.

Mr. Rajhans joined the Ontario Ministry of Health in July 1968, became head of the Dust Control Section one year later, and recently was named Chief of the Occupational Health Engineering Service. The Science Council of Canada appointed him as an expert consultant on the technical aspects of asbestos in 1976. As a registered professional engineer, Mr. Rajhans has served on the Industrial Ventilation and Air Sampling Committee of the American Conference of Governmental Industrial Hygienists since 1974. He has published numerous papers on the measurement and control of dusts.

Gordon M. Bragg is an associate professor of mechanical engineering at the University of Waterloo, Ontario, Canada. He received his BS in mechanical engineering from the University of Toronto in 1962 and his PhD in aeronautical engineering-fluid mechanics from the University of Cambridge in 1965. The author of numerous research papers in the fluid mechanics and environmental areas, he has also acted for some years as a consultant to industry and government in the fields of industrial ventilation and air pollution control.

PART I

ASBESTOS DUST: THE PROBLEM

In this section we discuss the origin of the dust, the dangers of being exposed to it, the methods of measuring it and the places one may expect to find it.

CHAPTER 1

INTRODUCTION

Asbestos fibers are used in many important industrial processes and are incorporated in a host of useful consumer products. Unfortunately, adverse health effects due to asbestos are well recognized and well documented and, hence, it would be imprudent to permit uncontrolled asbestos exposure. The purpose of this book is to provide information on the technical aspects of asbestos dust control to lead to better working and external environments and to contribute, thereby, to the limitation of asbestos health hazards.

The term "asbestos" does not describe a specific mineral; rather, it is used as a generic term for a group of hydrated mineral silicates which can appear in a fibrous crystal form (asbestiform), and, when crushed, can separate into flexible fibers. The most common types of asbestos fall into two mineral groups. These are the serpentine and amphibole groups. The only common type of asbestiform serpentine is chrysotile, which accounts for approximately 95% of the world's production and use. There are five types of amphibole asbestos of industrial significance. In decreasing order of quantity used they are amosite, crocidolite, anthophyllite, tremolite and actinolite. The forms and structures of asbestos are discussed in Chapter 3.

Asbestos dust enters the air as fine fibers or fibrils, and since the typical sizes are of micrometer dimensions, the dust falls slowly and can easily be inhaled. Typical dusts filtered from the air are shown in Figure 1.1. The particles or fibers have been collected on a Nuclepore filter and viewed through a scanning electron microscope. The circular black dots in the figure are holes in the surface of the filter, and the fibers are seen as whitish objects lying on the surface. The size of the fibers may be estimated since the smallest holes in the filter are approximately 0.2 μm in diameter.

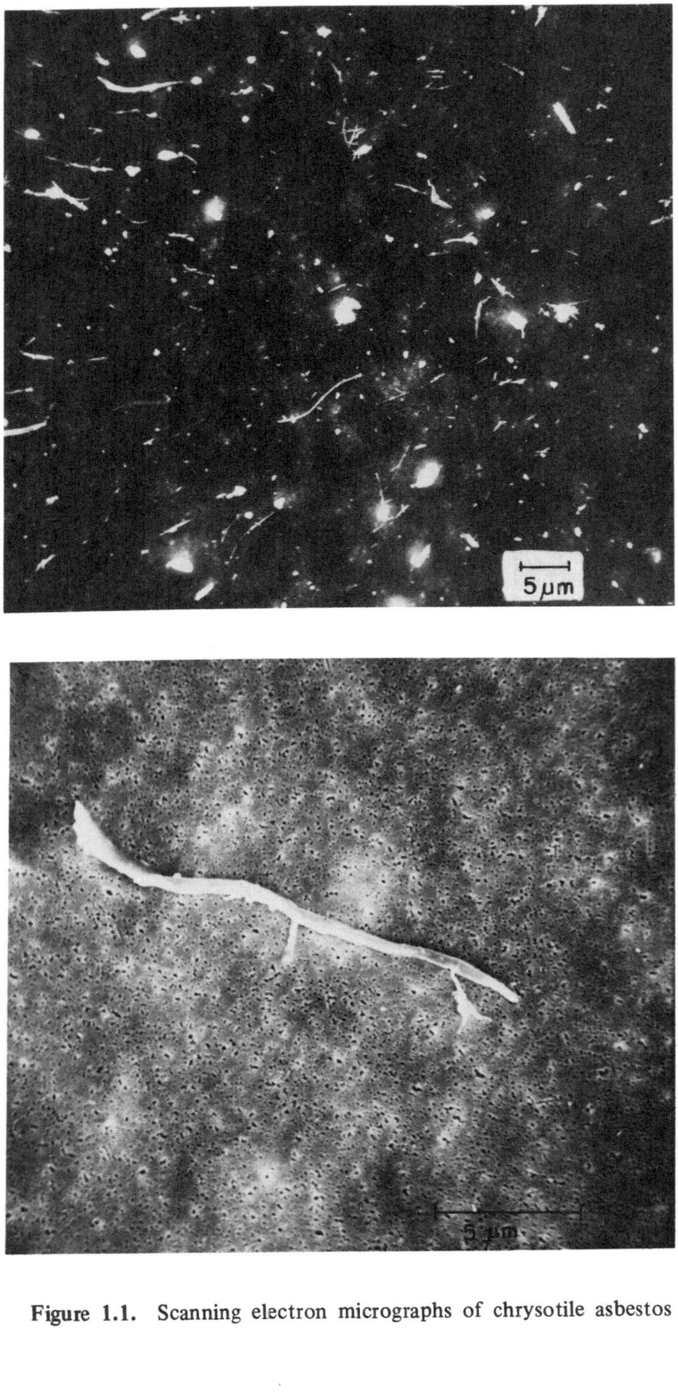

Figure 1.1. Scanning electron micrographs of chrysotile asbestos

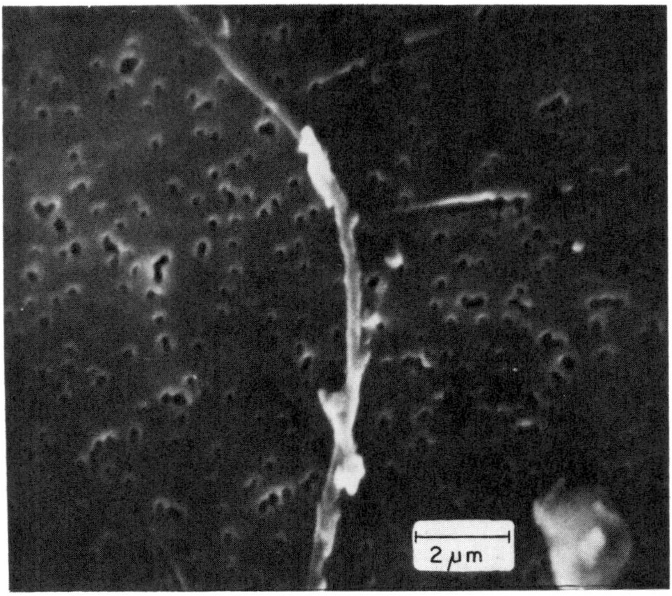

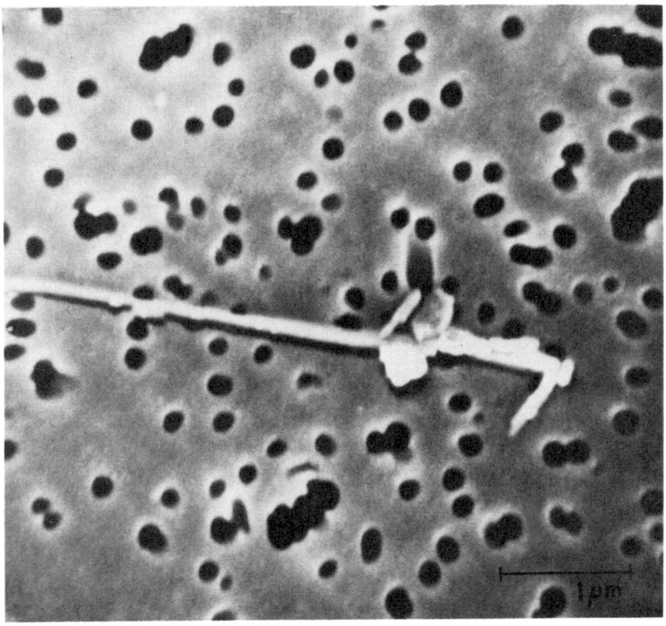

fibers filtered from the air onto a Nuclepore filter.

While the literature on the health aspects of asbestos is abundant, the references on the technical aspects of control are limited. For example, there are several review articles published on asbestos-related disorders by Parkes,[1] Enterline,[2] Meriweather and Price,[3] Advisory Committee on Asbestos Cancers,[4] Lewinsohn[5] and various others; however, there is no similar comprehensive review article on the technical aspects.

We shall first attempt to familiarize the reader with the various asbestos emission sources and exposure levels and to present an historical and critical review of the development of asbestos dust standards. We will then discuss the problem of sampling and evaluation techniques. With this background, the basic control methods may be conveniently presented. The rest of the book describes, in detail, the processes by which dispersions of asbestos fibers in the occupational and environmental areas can be reduced to allowable concentration levels.

REFERENCES

1. Parkes, W. R. "Asbestos Related Disorders," *Brit. J. Dis. Chest* 67:261-299 (1973).
2. Enterline, P. E. "Pitfalls in Epidemeological Research," *J. Occup. Med.* 18(3):150-156 (March 1976).
3. Meriweather, E. R. A., and C. W. Price. "Report on Effects of Asbestos Dust on the Lungs and Dust Suppression in the Asbestos Industry," Her Majesty's Stationery Office, London, 1930.
4. "Biological Effects of Asbestos," Report of the Advisory Committee on Asbestos Cancers to the Director of the International Agency for Research of Cancer, World Health Organization, International Agency for Research on Cancer, Lyon, France, October 5-6, 1972 *Annal. Occup. Hyg.* 16:9 (1973).
5. Lewinsohn, H. C. "Health Hazard of Asbestos: A Review of Recent Trends," *J. Soc. Occup. Med.* 24:2-10 (1974).

CHAPTER 2

THE ASBESTOS INDUSTRY

THE ASBESTOS INVENTORY

The worldwide asbestos industry mined and processed in 1977 approximately 6 million short ton of all kinds of asbestos. Approximately 45% of this was mined and used in iron curtain countries. Of the remainder, approximately 61% was produced in Canada, 15% in South Africa, 4% in the U.S. and 20% in other western countries.

Complete figures for all aspects of production are difficult to obtain for recent years. The data of Figure 2.1 show the flow of asbestos in the U.S. economy for 1973 in short tons.[1] The relative change in demand over a 10-year period is shown in Table 2.1. In most years, Canada provided approximately 95% of all imports to the United States. The small quantities from South Africa include crocidolite and amosite fibers.

Asbestos is graded by the producer and a typical range of Canadian grades is shown in Table 2.2 with prices per short ton as of January 1977.[2] The uses to which these grades are put may be seen in Table 2.3, which gives the distribution and use by grade in 1974.[1]

MINING AND MILLING

United States

The amount of mining in the United States is small compared to the market, as mentioned earlier. The mines in the United States are in California, Vermont, Arizona and North Carolina (Table 2.4).

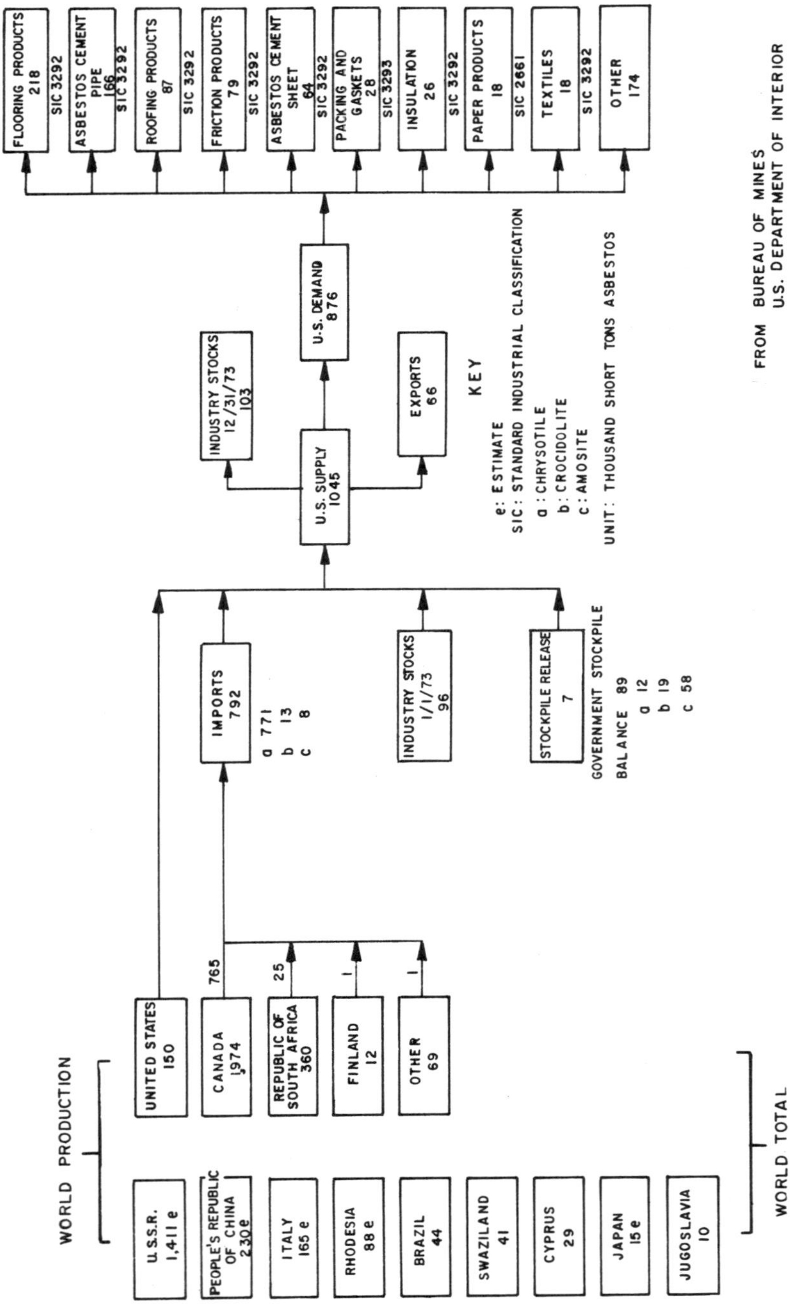

Figure 2.1. Supply/demand relationship for asbestos in the United States, 1973.

Table 2.1. United States–Asbestos Supply/Demand Relationships, 1965-1974 (Thousand Short Tons)

	1965	1966	1967	1968	1969	1970	1971	1972	1973	1974
World Mine Production										
United States	118	126	123	121	126	125	131	132	150	113
Rest of world	2,984	3,149	3,084	3,170	4,042	3,672	3,816	4,050	4,448	4,423
Total	3,102	3,275	3,207	3,291	4,168	3,797	3,947	4,182	4,598	4,536
Components of U.S. Supply										
Domestic mines	118	126	123	121	126	125	131	132	150	113
Government release	12	1	1	1	5	11	8	16	7	29
Imports, chrysotile	681	669	618	703	669	626	660	724	771	747
Imports, crocidolite	21	27	15	14	11	9	7	5	13	11
Imports, amosite	17	24	13	20	15	14	15	7	8	8
Industry stocks, Jan. 1	18	22	19	17	18	23	21	30	96	103
Total U.S. supply	867	869	789	876	844	808	842	914	1,045	1,011
Distribution of U.S. Supply										
Government acquisition	12	2	1	–	–	7	–	–	–	–
Industry stock, Dec. 31	17	15	20	18	24	20	29	46	103	103
Exports	43	47	47	41	36	47	54	59	66	62
Industry demand	795	805	721	817	784	734	759	809	876	846
U.S. Demand Pattern:										
Flooring products	200	202	179	204	196	184	191	202	218	153
Asbestos cement pipe	151	153	134	155	149	139	144	154	166	222
Roofing products	79	80	71	82	79	73	76	81	87	76
Friction products	71	72	65	74	71	66	68	73	79	80
Asbestos cement sheet	55	56	51	57	55	51	53	57	64	95
Packing and gaskets	24	24	22	25	24	22	23	24	26	29
Insulation	24	24	22	25	24	22	23	24	26	14
Paper products	16	16	14	16	16	15	15	16	18	63
Textiles	16	16	14	16	16	15	15	16	18	20
Other	159	162	145	163	154	147	151	162	174	94
Total U.S. demand	795	805	721	817	784	734	759	809	876	846

Table 2.2. Grades of Asbestos (Quebec) with Prices as of January 1977

Grade No.	Type	Price/Ton (Canadian $)
1	Crude	3,300-4,000
2	Crude	1,800-2,175
3	Spinning fiber	850-1,700
4	Asbestos cement fiber	464-950
5	Paper fiber	315-435
6	Paper and shingle fiber	230-275
7	Shorts	85-200

Canada

Mine production in Canada in 1974 was 1.8 million tons of chrysotile fiber having a value of $346 million. Canada's production of asbestos is derived from 14 mine operations, of which 10 are in Quebec (9 in the Thetford-Asbestos area of eastern Quebec and 1 in the northern Ungava region) and one each in Newfoundland, Ontario, British Columbia and Yukon Territory. Percentage distribution of production is as follows:

Quebec	–	82%
Yukon	–	7%
British Columbia	–	5%
Newfoundland	–	4%
Ontario	–	2%

Open pit mining represents 92% of the mine production and the remaining 8% is derived from underground operations. Mill capacity at these various mines varies from 3,000-33,000 ton/day of ore. Measured ore reserves in the larger mines are equal to 25-40 years of production, whereas in the smaller mines, the measured reserves amount to 10-12 years of operations. Considering the potential for additional ore in these same mines, it can be concluded that the main reserves of asbestos in the presently known mines can sustain the present Canadian production for nearly 50 years. The Canadian producers are listed in Table 2.5.

The asbestos ores mined contain an average of 6% of saleable fiber. An average of 95% of Canada's fiber production is exported. The remaining 5% fraction of roughly 75,000 tons is consumed in Canadian industry or stockpiled by producers.

Other

The production of South Africa is particularly important since most of the amphibole asbestos, *e.g.*, amosite and crocidolite, are mined here.

Table 2.3 United States–Asbestos Distribution by End Use, Grade and Type, 1974 (Short Tons)

	Chrysotile											
	Grades 1 and 2	Grade 3	Grade 4	Grade 5	Grade 6	Grade 7	Grade 8	Total Chrysotile	Crocido-lite	Amosite	Anthophyl-lite	Total Asbestos
Asbestos Cement Pipe	–	–	92,600	20,300	–	72,300	–	185,200	36,400	1,100	200	222,900
Asbestos Cement Sheet	–	–	10.900	15,800	52,000	11,800	–	90,500	–	4,300	–	94,800
Flooring Products	–	–	–	49,000	–	104,500	–	153,500	–	–	–	153,500
Roofing Products	–	–	3,700	11,100	12,700	46,300	–	73,800	–	1,700	–	75,500
Packing and Gaskets	100	1,900	7,500	7,500	1,400	10,300	–	28,700	100	–	–	28,800
Insulation: Thermal	–	100	900	100	3,400	2,800	–	7,300	–	1,800	–	9,100
Insulation: Electrical	–	–	400	1,600	–	2,700	–	4,700	–	–	–	4,700
Friction Products	–	5,600	1,300	29,500	6,300	36,600	300	79,600	–	–	200	79,800
Coatings and Compounds	–	–	100	400	300	36,700	400	37,900	–	–	–	37,900
Plastics	400	1,000	900	800	–	7,500	6,300	16,900	200	–	700	17,800
Textiles	700	14,200	3,900	800	–	800	–	20,400	–	–	–	20,400
Paper	–	–	5,500	400	23,900	33,300	–	63,100	200	–	–	63,300
Other	–	300	1,100	–	2,300	32,700	–	36,400	400	500	–	37,300
Total	1,200	23,100	128,800	137,300	102,300	398,300	7,000	798,000	37,300	9,400	1,100	845,800

Table 2.4. Asbestos Mines in the United States

	Location	Mining Plant
Arizona		
Asbestos Manufacturing Co.	Gila County	UG[a]
Jaquays Mining Corp.	Gila County	UG
Metate Asbestos Corp.	Gila County	UG
California		
Atlas Minerals Corp.	Fresno County	OP[b]
Calveras Asbestos Ltd.	Copperopolis	OP
Coalinga Asbestos Co.	Fresno County	OP
Pacific Asbestos Corp.	Calaveras County	OP
Union Carbide Corp.	San Benito County	OP
North Carolina		
Powhatan Mining Co.	Yancey County	OP
Vermont		
Vermont Asbestos Group	Orleans County	OP

[a] UG: underground.
[b] OP: open pit.

South Africa is the only commercial source of amosite. Rhodesian chrysotile is of spinning grade and, hence, is also in high demand. The 1973 production of asbestos by type is shown in Table 2.6.

The Milling Process

The recovery of asbestos minerals from host rocks begins with mine extraction methods (mainly open pit) similar to those employed in all other mines. On its arrival at the mill plant, the asbestos ore is subjected to a specific dry treatment process, which uses air to effect the required mechanical separations leading to fiber recovery. The fiber extraction process is thus composed of four successive basic steps (Figure 2.2):[3]

1. *Ore Preparation:* crushing, sizing, sorting and drying
2. *Rock Circuits or Lines:* sequence of impact crushing, fiberizing and air-lifting of asbestos on shaking screens
3. *Fiber Circuits or Lines:* air-lifted fiber from "rock lines" is dedusted in cyclones, fiberized, sized and cleaned (and mixed) ready for packaging
4. *Dust Treatment:* further screening, pulverizing and sizing of all dust in the bagroom to recover saleable short (float-grade) fibers

To produce each ton of fiber, eight tons of air are used to effect the various separation functions and two tons are used to carry out the dust control functions. The metallurgical balance of the recovery process, of the total ore milled is:

40% is rejected as coarse rock and sand-sized fragments
53% is rejected as fine rock fragments and powders
5% is recovered as fiber (3%–grades 1 to 6; 2%–grade 7)
2% is recovered as float grades (dust recovery)

100%

Table 2.5. Chrysotile Asbestos Mines and Mills in Canada

Producer	Mine Location	Mining Plant	Mill Capacity (ton/day)
Newfoundland			
Advocate Mines Ltd.	Baie Verte	OP[a]	7,000
Quebec			
Asbestos Corp., Ltd. Thetford Mines, Quebec	British Canadian Mine, Black Lake Mill #1	OP	6,600
			7,200
	Mill #2	OP	7,000
	King-Beaver Mine, Thetford Mines	UG,[b] OP	–
	Normandie Mine, Normandie	OP	8,000
	Asbestos Hill Mine, Putuniq	OP	5,500
Bell Asbestos Mines Ltd.	Thetford Mines, Thetford Mines, Que.	UG, OP	3,000
Canadian Johns-Manville Co., Ltd.	Jeffrey Mine, Asbestos	OP	33,000
Carey-Canadian Mines Ltd.	East Broughton	OP	6,000
Lake Asbestos of Quebec, Ltd.	Black Lake	OP	9,000
National Mines Division	Thetford Mines	OP	3,500
Ontario			
Hedman Mines Ltd.	Matheson	OP	300
United Asbestos Inc.	Matachewan	OP	3,500
British Columbia			
Cassiar Asbestos Corp., Ltd.	Cassiar Mine, Cassiar	OP	4,000
Yukon Territory			
Cassiar Asbestos Corp., Ltd.	Clinton Mine, Clinton Creek	OP	5,300

[a] OP: open pit.
[b] UG: underground.

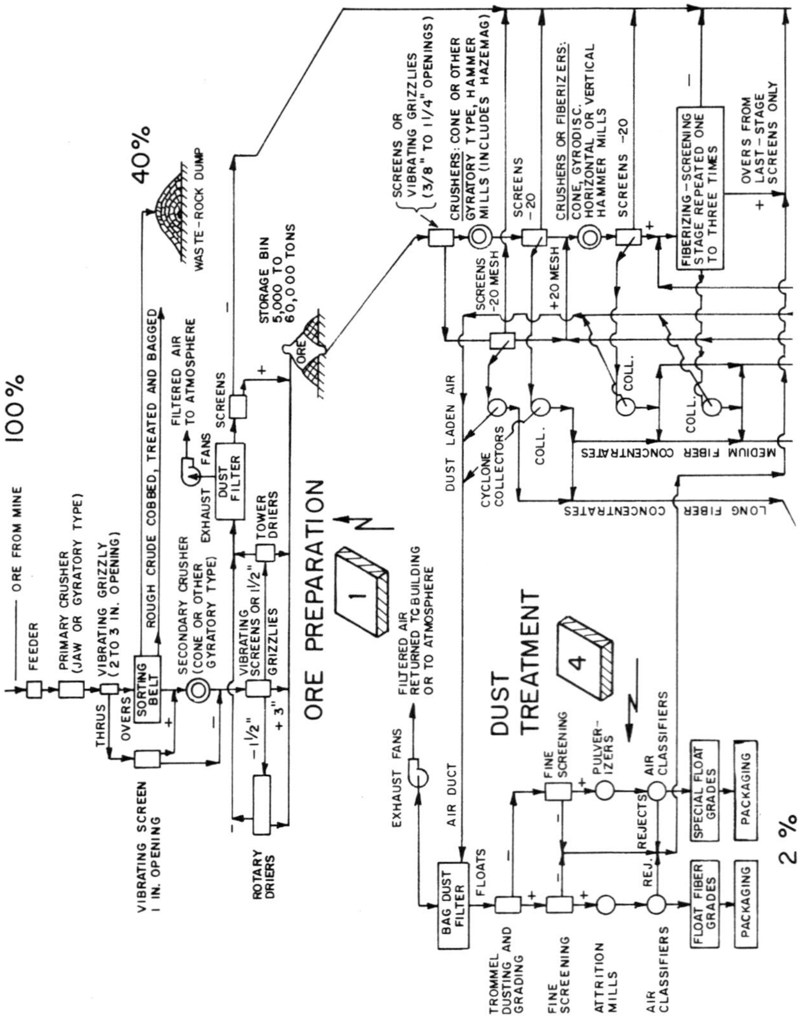
ORE FROM MINE
100%
FEEDER
PRIMARY CRUSHER (JAW OR GYRATORY TYPE)
VIBRATING GRIZZLY (2 TO 3 IN. OPENING)
THRUS
OVERS
SORTING BELT
VIBRATING SCREEN 1 IN. OPENING
ROUGH CRUDE COBBED, TREATED AND BAGGED
WASTE-ROCK DUMP
40%
SECONDARY CRUSHER (CONE OR OTHER GYRATORY TYPE)
VIBRATING SCREENS OR 1 1/2" GRIZZLIES
-1 1/2"
+3"
ROTARY DRIERS
TOWER DRIERS
DUST FILTER
EXHAUST FANS
FILTERED AIR TO ATMOSPHERE
SCREENS
STORAGE BIN 5,000 TO 60,000 TONS
ORE
ORE PREPARATION
1
SCREENS OR VIBRATING GRIZZLIES (3/8" TO 1 1/4" OPENINGS)
CRUSHERS: CONE OR OTHER GYRATORY TYPE, HAMMER MILLS (INCLUDES HAZEMAG)
SCREENS -20
CRUSHERS OR FIBERIZERS: CONE, GYRODISC. HORIZONTAL OR VERTICAL HAMMER MILLS
SCREENS -20
SCREENS -20 MESH
+20 MESH
FIBERIZING-SCREENING STAGE REPEATED ONE TO THREE TIMES
OVERS FROM LAST-STAGE SCREENS ONLY
DUST LADEN AIR
CYCLONE COLLECTORS
COLL.
MEDIUM FIBER CONCENTRATES
LONG FIBER CONCENTRATES
EXHAUST FANS
FILTERED AIR RETURNED TO BUILDING OR TO ATMOSPHERE
AIR DUCT
BAG DUST FILTER
FLOATS
DUST TREATMENT
4
TROMMEL DUSTING AND GRADING
FINE SCREENING
ATTRITION MILLS
AIR CLASSIFIERS
PULVER-IZERS
REJECTS
REJ.
FLOAT FIBER GRADES
SPECIAL FLOAT GRADES
PACKAGING
2%

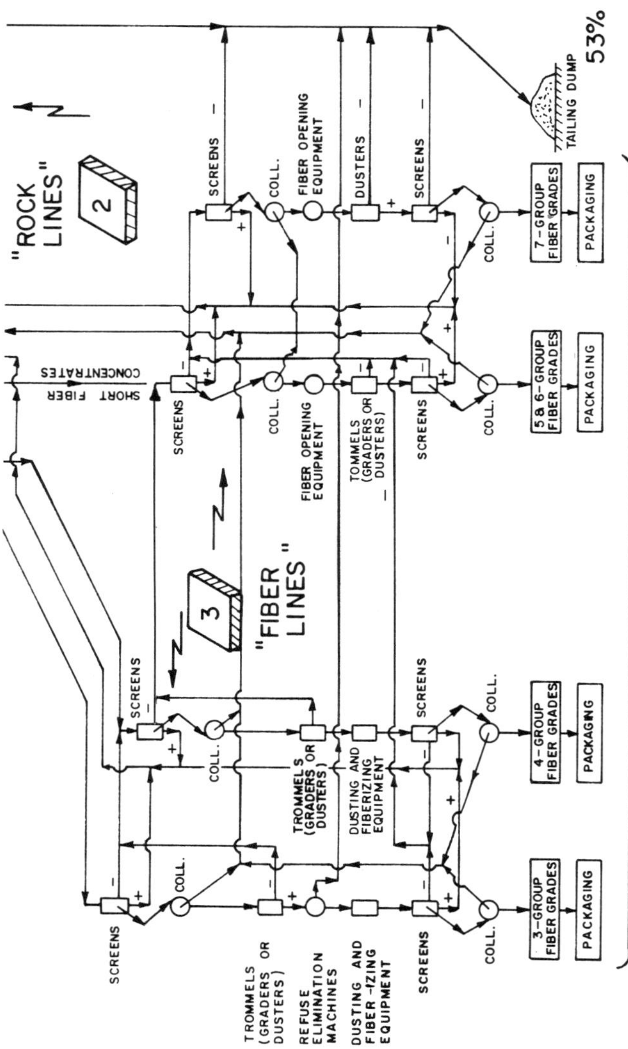

Figure 2.2. Generalized flow sheet of Canadian asbestos mill.

Table 2.6. World Asbestos Production, 1973 (Thousands of Short Tons)

	Production (1973)
Chrysotile	
North America	
Canada	1,974
Mexico	–
United States	150
Total	2,124
South America	
Argentina	1
Brazil	44
Total	45
Europe	
Bulgaria	3
Finland	12
Italy	165
U.S.S.R.	1,411
Yugoslavia	10
Other	3
Total	1,604
Africa	
Rhodesia	88
South Africa, Republic of	75
Swaziland	41
Other	1
Total	205
Asia	
China, People's Republic of	230
Cyprus	29
India	–
Japan	15
Korea, Republic of	6
Taiwan	3
Turkey	5
Total	300
Oceania	
Australia	35
World Chrysotile Total	4,313
Crocidolite	
South Africa, Republic of	168
Amosite	
South Africa, Republic of	117

MANUFACTURING INDUSTRIES

Industrial Uses

Asbestos has been a known health hazard for almost 35 years, and yet during this same period, on a worldwide basis, the use of asbestos has risen at an accelerating rate. The basic reason is that asbestos has astonishing physical properties. It can withstand heat exceeding 500°C and is resistant to the actions of acids and many other chemicals. It also has excellent absorption and filtering properties. Due to its extreme flexilibity, asbestos fibrils can be carded, spun and woven as easily as cotton or jute fibers; yet, each has a high tensile strength approximately equal to that of steel. There are well over 3,000 recorded uses of asbestos, and these are so wide-ranging that only a broad outline can be given. A summary of of some of the more common uses is given in Table 2.7.

Asbestos fibers can be classified broadly into longer and shorter grades. Use of a special grade not only depends upon application, but also on price and availability. However, longer-grade fibers are normally used to make textile products because they can be easily carded, spun and woven into yarn, rovings, tape and cloth. The textiles containing asbestos are used to make fireproof clothing, packings, gaskets, lagging cloth and laps that provide electrical and thermal insulation. Shorter fibers are used mostly in asbestos-cement products including pipe, roofing shingles, flat and corrugated siding, millboard, cements and fillers. These account for well over half the total use of asbestos. Short chrysotile fibers are also used in the manufacture of friction materials, *e.g.*, brake linings and clutch facings and automatic transmission linings.

Asbestos products can be characterized according to the susceptibility of the asbestos to removal from the product by mechanical means. We can term these "locked-in" and "free" materials. The "locked-in materials" are plastic floor tiles, paints, boat hulls, etc. The examples of "nonlocked-in materials" are gaskets, acoustic tiles, thermal insulation and domestic irons. For thermal insulation, both crocidolite and chrysotile have been used for pipe and boiler lagging, for grouting expansion gaps in furnaces and kilns and as spray consisting of 5-20% chrysotile with clay binders and synthetic resins for thermal and acoustic insulation of buildings and ships.

The uses and applications of asbestos fibers also depend on their characteristics. For example, chrysotile, because of its reinforcing quality, is commonly used in the asbestos-cement industry. Amosite, because of its heat resistance, is used in insulation and building products. Crocidolite, due to its excellent acid resistance, has been used in the manufacture of battery boxes and packings for acid pumps. The friction materials, including

Table 2.7. Common Uses of Asbestos[4]

Asbestos Friction Materials
- Brake Linings
  - Woven, containing asbestos yarn, tape or cloth
  - Molded, including all nonwoven types
- Clutch Facing
  - Woven, containing asbestos yarn, tape or cloth
  - Nonwoven–clutch lining; transmission lining

Asbestos-Cement Materials
- Flat sheets and wallboard, all thicknesses converted to ¼-in. basis
- Slabs
- Corrugated sheets
- Pipes
- Corrugated tiles for roofs in industry, agriculture and dwellings
- Planks for platforms in buildings under construction
- Balcony canopies
- Rain gutters
- Interior walls
- Ventilating shafts
- Air conditioning assemblies
- Pressure piping (for underground drinking water distribution systems, fuel gas and sewage)
- Cooling towers (electricity-generating stations)
- Asbestos felts
  - Roofing-asphalt or tar saturated
  - Other
- Other asbestos and asbestos-cement products, including millboard and prefabricated housing components

Asbestos Shingles and Clapboard
- Siding shingles and clapboard, including accessories
- Roofing shingles

Asbestos Textiles
- Yarn, cord and thread
- Cloth
- Other asbestos textiles, including roving, lap, wick, rope, tape, carded fibers, etc.

Gaskets and Insulation
- Gaskets, all types
- Gaskets (for sealing nonmoving parts)
  - Asbestos, asbestos-metallic and asbestos-rubber
- Packing (except leather, rubber and metal) and asbestos insulations
  - Asbestos compressed sheet
- Packing (for sealing moving parts)
  - Asbestos, asbestos-metallic and asbestos-rubber
- Insulation materials containing asbestos pipe insulation (cellular and laminated)
  - 85% magnesia
  - Diatomaceous silica, calcium silicate, expanded silica and asbestos fiber

Table 2.7., continued

Other block insulation, including cellular and laminated
All other asbestos insulation
Asbestos Boards and Papers
Boards
Filtering and clarifying
Coverings, coatings, casings and jacketings for all kinds of surfaces
Manufacturing of welders' and melters' shields
Slideways in the glass industry
Handles and fire doors
Auto parts
Safes
Protective walls
Curtains, etc.
Sheets
Inner/outer linings of furnaces and heating vessels, drying ovens, incubators, heaters, climate-controlled spaces, etc.
Plates
Insulating buildings against vibrations (aluminum-asbestos)
Solar-heat reflecting surfaces (70% of solar heat)
Special Asbestos Papers
Filters
Asbestos Plastics
Flooring tiles (asbestos-asphalt tiles and, increasingly, asbestos-polymers of vinyl)
Pressed or molded (thermal insulation and in electrical machinery)
Resinated asbestos felt (manufacturing of wings and firing of missiles and expansion cones for nozzles of boost motors). Other uses in aircraft industry: nozzles for motor tubes, missile tailpipes and missile-heat barriers, fuselages for guided missiles, fuel tanks for fighter bombers, cabin floors, etc.
Asbestos Acid-Resistant Compositions
Used mostly in chemical industry

disc brake pads, passenger car and commerical vehicle brake linings and clutch facings, contain more than 50% chrysotile. Chrysotile is used in friction materials:

1. because of its heat stability
2. as a reinforcing agent
3. as a filler
4. as a regulator or inhibitor of the resin flow in moulding
5. as a dispersing agent for the metal chips and other particles during formation
6. because it is less abrasive than other heat-stable fillers

Crocidolite and amosite fibers, when treated with inorganic binders, can withstand temperatures higher than 250°C. Thus, they can be made into cloth, packings, yarns and baggings, which show good chemical and thermal resistance. Because amosite contains air pockets, heat insulation is further increased.

Anthophyllite asbestos has good heat and acid resistance. It has been treated with resins and used in plastics. Although this fiber does not have the strength of chrysotile, crocidolite or amosite, it has good bulking properties. Some anthophyllites are mostly iron-free, almost white in color and are desirable in some types of plastics.

Tremolite, after an acid and caustic wash, is considered a good fiber for use in the chemical laboratory for filtering chemicals and is sold for that purpose. However, because of its relative abundance, chrysotile is used wherever possible in preference to all other forms of asbestos.

It is not possible here to explore fully all the manufacturing processes in which asbestos is used. In addition, these processes are constantly being altered and improved in modern plants. While some examples of the more common processes are given in Chapter 7, we shall concentrate our control methods on single elements of the complete process. In manufacturing, certain elemental processes tend to cause more dust than others. Some examples of dust-causing processes (in arbitrary order) are:

crushing	screening	drying
storage	debagging	mixing
grinding	milling	drilling
punching	sanding	polishing
carding	spinning	twisting
braiding	spraying	

Many other processes are possible, but the above list contains a large proportion of the dust-causing operations in most plants. These processes and their control will be discussed in Chapters 7, 8, 10 and 11.

Manufacturing and processing of asbestos is concentrated in the industrial regions of the United States and Canada, and, as a result, the largest number of plants are in New York, Pennsylvania, Ohio, Michigan, Illinois and California. In Canada, processing is concentrated mostly in Ontario, Quebec and British Columbia. The major processors in the U.S. have been documented extensively by the Environmental Protection Agency[5] in the United States and by the Health and Welfare Department in Canada[6] and this information is summarized in Appendix 1.

REFERENCES

1. Interior, U.S. Department of. "Asbestos," in *Mineral Facts and Problems,* Bulletin 667, U.S. Government Printing Office (1975.)

2. Interior, U.S. Department of. "Asbestos," in *Mineral Facts and Problems, Asbestos,* (February 1977) p.32
3. Jenkins, G.F. "Asbestos," in *Industrial Minerals and Rocks,* American Institute of Mining Engineers (1960).
4. Ontario Ministry of the Environment. "Asbestos as a Hazardous Contaminant II" (January 1975).
5. Environmental Protection Agency. "National Inventory of Sources and Emissions - Asbestos" 1968, APTD-70, National Technical Information Service, Springfield, VA (1970).
6. Health and Welfare Canada. "Report of the Asbestos Working Group," Queen's Printer, Ottawa (1976).

CHAPTER 3

MINERALOGY AND PHYSICAL/CHEMICAL PROPERTIES OF ASBESTOS

MINERALOGY OF ASBESTOS

As mentioned earlier, asbestos is a generic term applied to fibrous silicate minerals. Six common asbestos varieties belong either to the serpentine group or to the amphibole group of minerals. The amphibole varieties are five in number – amosite, anthophyllite, crocidolite, tremolite and actinolite. The single serpentine variety, chrysotile, accounts for more than 90% of all asbestos uses.

Origin

Chrysotile occurs as veins in serpentine-rich rocks. The compact fibers in the veins are most often oriented perpendicularly to the vein walls in a regular fashion and simply represent serpentine recrystallization along fractures in the host serpentine rock. The amphibole varieties occur as irregular fibrous aggregates in high-temperature metamorphic schists.

Crystal Structure

Although finely fibrous in habit, chrysotile belongs to a class of sheeted crystal structures. The sheets are composed of alternating layers of SiO_2 tetrahedra bound together by hydroxyl (OH) and magnesium ions, as shown in Figure 3.1. The fibrous form results from the growth pattern of the chrysotile sheet structure giving rise to spirally wound tubes, as in Figure 3.2. This fibrous growth pattern has significant implications with regard to mineralogical identification and analyses in that, obvious as the fibers may appear to the eye, mineralogical identification and measurement based on any of the usual indirect analytical methods is practically

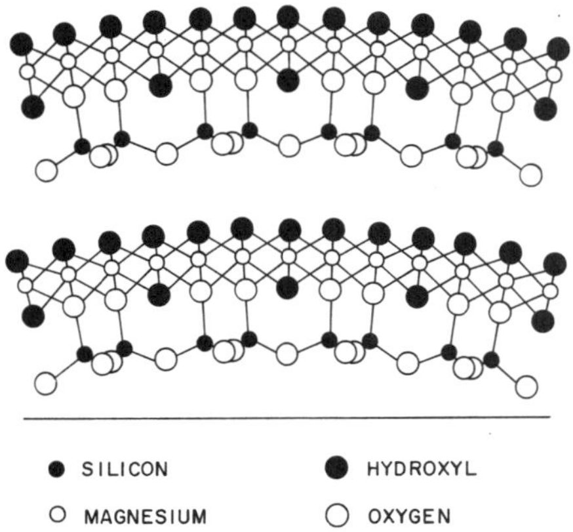

Figure 3.1. Structure of chrysotile asbestos crystals.

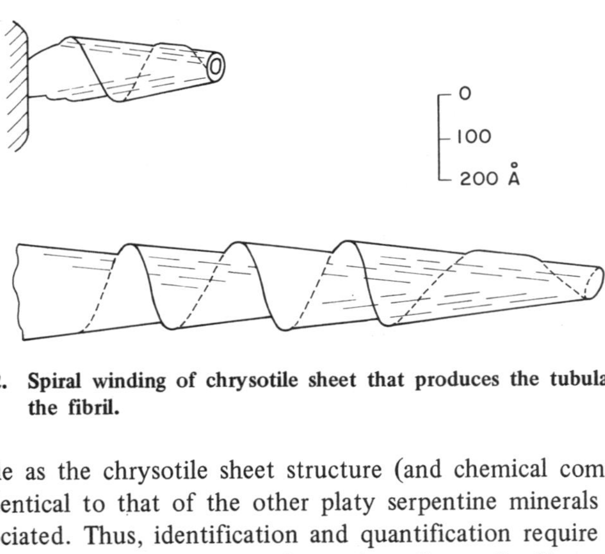

Figure 3.2. Spiral winding of chrysotile sheet that produces the tubular shape of the fibril.

impossible as the chrysotile sheet structure (and chemical composition) is nearly identical to that of the other platy serpentine minerals with which it is associated. Thus, identification and quantification require both visual and indirect techniques to be used together. Chrysotile fibers are very small in diameter, tubular and very soft and flexible. Individual fibrils measure 100-250 Å in diameter.

Amphiboles, on the other hand, have a fibrous habit, which is derived from a chain-like crystal structure. The fibers are larger in diameter, straight and solid in nature and, thus, hard and springy. Because of structural differences (and chemical differences) within the amphibole groups, identifications and measurements of amphibole asbestos minerals are somewhat easier than is the case for chrysotile. Yet, the fibrous nature presents special problems for most indirect analytical techniques, and visual assessment or control is usually carried out in conjunction with indirect techniques of analysis. Additional pertinent mineralogical parameters are indicated in Table 3.1.[1]

PHYSICAL AND CHEMICAL PROPERTIES OF CHRYSOTILE

The significant physical and chemical properties of chrysotile are summarized in Table 3.1. The mineral is a hydrous magnesium silicate $Mg_3Si_2O_5(OH)$. The normal mineral impurities are magnetite, brucite, magnesite and pyrrhotite. Thus, the important chemical impurities are aluminum, iron, chrome and nickel. Acids will easily attack chrysotile.

On heating above 300°C, chrysotile begins to loose its water of crystallization and the fibers begin to lose their flexibility. At 580°C, the fibrous nature is destroyed and the chrysotile transformed to nonfibrous magnesium olivene.

PHYSICAL AND CHEMICAL PROPERTIES OF AMPHIBOLE ASBESTOS

The principal properties of these fibrous minerals (crocidolite, amosite, anthophyllite, tremolite and actinolite) are also listed in Table 3.1. As a group, their composition is that of a hydrated alkali iron-magnesium silicate. Chemical compositions are, however, complex and variable, due to ionic replacements. Mineral impurities are quartz, magnetite and hematite. Amphibole asbestos is typically resistant to acid.

Physically, these fibers are generally more refractory than chrysotile; loss of water and fiber deterioration thus occurs at somewhat higher temperatures (~1,000°C).

PROPERTIES OF ASBESTOS IN RELATION TO BIOLOGICAL EFFECTS

There does not appear to be any general medical consensus which would relate the various asbestos health disorders to specific properties or parameters of the asbestos fibers. However, the various possible causes that have been suggested can be grouped into three main categories, each of which would emphasize the toxic effects of particular properties:

1. Toxic effects are mainly physical and mechanical in nature and related to the size and shape of the respirable fibers.
2. Toxic effects are mainly chemical in nature and related to fiber dissolution and liberation of toxic elements – Mg, Ni, Cr.

Table 3-1. Physical, Chemical and Mineralogical Properties of Varieties of Asbestos[1]

Property	Chrysotile	Crocidolite	Amosite	Anthophyllite	Tremolite	Actinolite
Chemical Formula	$Mg_3Si_2O_5(OH)_4$	$Na_2Fe_5Si_8O_{22}(OH)_2$	$(FeMg)_6Si_8O_{22}(OH)_2$	$(FeMg)_7Si_8O_{22}(OH)_2$	$Ca_2Mg_5Si_8O_{22}(OH)_2$	$(CaMgFe)_6Si_8O_{22}(OH)_2$
Essential Composition	Hydrous silicate of magnesia	Silicate of sodium and iron with some water	Silicate of iron and magnesium; higher iron than anthophyllite	Magnesium silicate with iron	Calcium and magnesium silicate with some water	Calcium-magnesium-iron silicate; water up to 5%
Percentage Chemical Composition						
SiO_2	37-44	49-53	49-53	56-58	51-62	–
MgO	39-44	0-3	1-7	28-34	0-30	–
FeO	0.0-6.0	13-20	34-44	3-12	1.5-5.0	–
Fe_2O_3	0.1-5.0	17-20	–	–	–	–
Al_2O_3	0.2-1.5	–	2-9	0.5-1.5	1.0-4.0	–
H_2O	12-15	2.5-4.5	2-5	1.0-6.0	0-5.0	–
CaO	trace to 5.0	–	–	–	0-18	–
Na_2O	–	4.0-8.5	–	–	0-9	–
CaO and Na_2O	–	–	0.5-2.5	–	–	–
pH	9.2-9.8	–	–	Neutral	–	–
Resistance to Acids	Poor	Good	–	–	Good	Good
Veining	Cross and slip fibers	Cross fiber	Cross fiber	Slip, mass fiber unoriented and interlacing	Slip or mass fiber	Slip or mass fiber
Color	Green, gray, amber to white	Blue	Gray, yellow to dark brown	Yellowish brown, grayish white	Gray-white, greenish-yellowish, bluish	Greenish

Texture	Soft to harsh, also silky	Soft to harsh	Coarse but somewhat pliable	Harsh	Generally harsh, sometimes soft	Harsh
Luster	Silky	Silky to dull	Vitreous, somewhat pearly	Vitreous to pearly	Silky	Silky
Hardness[a]	2.5-4.0	4	5.5-6.0	5.5-6.0	5.5	6±
Flexibility	High	Good	Good	Poor	Poor	Poor
Spinnability	Very good	Fair	Fair	Poor	Poor	Poor
Tensile Strength, lb/in.^{2}	824,000 max.	876,000 max.	16,000-90,000	4,000 and less	1,000-8,000	1,000 and less
Fusion Point, °F	2,770	2,180	2,550	2,675	2,400	2,540
Specific Heat, Btu/lb-°F	0.266	0.201	0.193	0.210	0.212	0.217
Electric Charge	Positive	Negative	Negative	Negative	Negative	Negative
Filtration Properties	Slow	Fast	Fast	Medium	Medium	Medium
Specific Gravity	2.4-2.6	3.2-3.3	3.1-3.25	2.85-3.1	2.9-3.2	3.0-3.2
Cleavage	010 perfect	110 perfect	110 perfect	110 perfect	110 perfect	110 perfect
Optical Properties	Biaxial positive, extinction parallel	Biaxial ±, extinction inclined	Biaxial positive, extinction parallel	Biaxial positive, extinction parallel	Biaxial negative, extinction inclined	Biaxial negative, extinction inclined
Refractive Index	1.50-1.55	1.7 pleochroic	1.64 ±	1.61 ±	1.61 ±	1.63 ±, weakly pleochroic
Resistance to Destruction by Heat	Good, brittle at high temperatures	Poor, fuses	Good, brittle at high temperatures	Very good	Fair to good	–
Temperature at Ignition Loss, °F	1,800	1,200	1,600-1,800	1,800	1,800	–
Magnetite Content, %	0.0-0.5	3.0-5.9	0	0	0	–

Table 3.1, Continued

Property	Chrysotile	Crocidolite	Amosite	Anthophyllite	Tremolite	Actinolite
Crystal Structure	Fibrous and asbestiform	Fibrous	Prismatic, lamellar to fibrous	Prismatic, lamellar to fibrous	Long and thin columnar to fibrous	Long and thin columnar to fibrous
Crystal System	Monoclinic and orthorhombic	Monoclinic	Monoclinic	Orthorhombic	Monoclinic	Monoclinic
Mineralogical Structure	In veins of serpentine, etc.	Fibrous in iron stones	Lamellar, coarse to fine fibrous and asbestiform	Lamellar, fibrous asbestiform	Long, prismatic and fibrous aggregates	Reticulated long prismatic crystals and fibers
Mineral Association	In altered peridotite adjacent to serpentine and limestone near contact with basic igneous rocks	Iron-rich silicious argillite in quartzose schists	In crystalline schists, etc.	In crystalline schists and gneisses	In Mg limestones as alteration product of magnesian rocks, metamorphic and igneous rocks	In limestones and in crystalline schists

[a]Working Scale of Hardness: 1, very easily scratched by fingernail, and has greasy feel to the hand; 2, easily scratched by fingernail; 3, scratch by brass pin or copper coin; 4, easily scratched by knife; 5, scratch with difficulty with knife; 6, easily scratched by file; 7, little touched by file, but will scratch window glass. All harder than 7 will scratch window glass.

3. Toxic effects are physico-chemical in nature and related to surface characteristics of the fiber as is evidenced by the reaction of chrysotile in water and biological fluids.

That different explanations such as these are suggested, is not unexpected. Part of the reason lies in the fact that there are known to be a number of different asbestos-related health disorders. In fact, there is a consensus regarding asbestosis, which is deemed to be related to mechanical effects of particular fiber sizes (>5 μm long and < 5 μm in diameter).

It can be appreciated that the identification of parameters directly responsible for disorders has a strong bearing on the nature of the standards of exposure that should be promulgated.

REFERENCES

1. Environmental Protection Agency. "Control Techniques for Asbestos Air Pollutants", Publ. AP-117, EPA, Research Triangle Park, NC (February 1973.)

CHAPTER 4

ASBESTOS EMISSION SOURCES

This chapter deals with the various sources responsible for the emission of asbestos fibers into the occupational and nonoccupational environments.

NATURAL EMISSIONS

Natural phenomena and nonindustrial activities can contribute in a small way to local asbestos fiber emissions into the environment. Weathering and erosion of outcrops of asbestos-containing rocks will liberate a small amount of fiber. Construction activity (foundations, excavations) in similar-type rocks will also give rise to asbestos emissions. Farming of soils developed over asbestos-containing rocks will also disseminate fibers.

Extensive data on ambient or background levels of asbestos fibers, and variation with geographic location and seasonal variation are not available. However, total emission in the U.S. to the atmosphere has been estimated at 54 tons per year.[1] Data taken by the authors in regions where there is no significant mining or manufacturing suggest that either on a mass basis or on a fiber-counting basis the natural background level is less than 1% of the standards used in occupational work; however, background levels in cities can be expected to be much higher.[2] This is, of course, because the background levels are due to nonnatural sources.

MINING AND MILLING

The various processes of asbestos mining and milling are significant sources of fiber emission. The very nature of these processes, which are those of blasting, rock extraction, crushing and grinding, and transportation of ore for open pit mining are very prone to fiber emission, especially if they are carried out without proper control methods. In addition, the tailings heaps are exposed to wind and to disturbances by bulldozing. Road surfaces constructed of asbestos-containing overburden are also potential emission sources. Fibers are also emitted during drying, crushing, grinding and screening of the ore. The shipment of milled asbestos fiber, usually in bags, can result in emissions due to broken bags and spillage. Asbestos-containing solid wastes when transported in open vehicles through urban areas, can be a significant emission source.

MANUFACTURING PROCESSES

The handling of asbestos prior to the manufacture of asbestos products constitutes potential emissions of fiber that are common to most asbestos-using industries. These include such functions as unloading and storage of fibers, bag opening and dumping, fluffing and mixing processes. These processes generate by far the highest emission levels, especially when carried out in the dry state.

The wet manufacturing processes for asbestos cement products, floor tiles, paper products, friction materials etc. do not emit a significant amount of fibers. However, the sawing, turning, grinding, buffing and other finishing steps may create an excessive amount of fibrous emissions. There are unusual potential emission sources in asbestos textile manufacturing industries, such as opening, carding, spinning and weaving. Asbestos fibers removed by local ventilation and filtering devices and not recirculated into the production process are also sources of emission.

PRODUCT USE

The uses of asbestos-containing products have long been potential sources of asbestos emissions. The most recognized ones are spraying of asbestos insulation materials, the use of friction products, fireproofing materials in construction and the use of asbestos asphalt pavements. The emissions created by these uses are more noticeable in urban areas than elsewhere.

The asbestos fibers used in asbestos cement products, shingles and floor tiles are tightly bound within the products and undergo little actual abrasion or wear before being discarded.

ESTIMATES OF EMISSION QUANTITIES

Very few attempts have been made to estimate the quantities of fiber emissions from the sources discussed above. However, the following three tables should give some indicated of such estimates. Table 4.1[3] shows a

Table 4.1. Estimates of the Nationwide Emission Quantities from Various U.S. Industries

				Emissions if uncontrolled		Emission with Present Degree of Control		Emissions with Best Degree of Control	
	Tons/yr (1969)	Percentage of Total Consumption	Number of Plants	Emission Factor (lbs/ton)	Emission (tons/yr)	Control (%)	Emission (tons/yr)	Best Control (%)	Emission ton/yr
Mining (Domestic)	127,000			10	635	50	317	80	121
Milling (Domestic)	127,000		9	100	6,350	80	1,210	99	63.5
Imported Fiber	695,000								
Total Fiber	822,000								
Processing									
Asbestos cement	412,500	50		4	825	75	206	99	8.3
Floor tile	201,200	24.6		4	402	75	101	99	4.0
Friction products	104,600	12.8	361	120	6,276	95	314	99	63
Textile	18,100	2.1		40	362	95	18	99	3.6
Paper	30,200	3.8		4	60	75	15	99	0.6
Miscellaneous	6,800	0.8		10	34	75	9	99	0.3
Consumptive Uses									
Construction industry	5,125	0.6		25	64	N/C	64	90	6.4
Brake linings	38,400	4.7	several	20	384	N/C	384	90	38.4
Steel fireproofing	3,050	0.4	thousands	40	61	75	15	90	6.1
Insulating cement	2,025	0.25		40	41	75	10	90	4
		Total Emissions (tons/yr)			15,500		2,663		319.2

set of estimates of emissions based on a series of estimates of collection efficiency of existing equipment. Table 4.2[4] gives another set of estimates of emissions from varying processes and Table 4.3[4] gives the estimates for mining and milling processes.

Table 4.2. Emission Factors for Processing of Asbestos[a]

Source[b]	Emission Factor Asbestos Processed (lb/ton)
Friction Material, Controlled	6
Asbestos Cement Products, Controlled	1
Textiles	40
Textile, Controlled	2
Asbestos Paper	4
Asbestos Paper, Controlled	1
Floor Tile	4
Floor Tile, Controlled	1
Consumptive Uses of Asbestos[a]	
Brake Linings	10
Steel Fireproofing, Controlled	10
Insulating Cement, Controlled	25
Construction Industry	25

[a]The emission factors are based on engineering estimates and not source sampling. These factors cannot be used to quantify asbestos emissions.
[b]Uncontrolled, unless otherwise specified.

Table 4.3. Emission Factors for Asbestos from Mining and Milling[a]

Source[b]	Emission Factor Asbestos Produced (lb/ton)
Mining, Total	9-10
Mining	3
Loading	2
Hauling	2
Unloading	2
Mining, Total, 50% Control	5
Milling	100
Milling, 50% Control	80
Milling, 80% Control	20

The main conclusions to be drawn from these are that:

1. although the amount of emissions is given in absolute terms, the amounts are only engineering estimates and cannot be considered to have great accuracy;
2. the tables show the relative emission rates of various sources; for example, the highest emission rates are from milling, friction products and mining, respectively; and
3. the emission rates can be significantly reduced by applying appropriate and adequate control measures. As can be seen, the reduction in emissions can be of the order of 10- to 20-fold by applying good engineering practice.

REFERENCES

1. Environmental Protection Agency. "Hazardous Wastes," EPA-600/5-77-002, National Technological Information Centre (February 1977).
2. Nicholson, W.J. "Measurement of Asbestos in the Ambient Air," Report to the Environmental Protection Agency, Contract CPA70-92 (1971).
3. State of Illinois Institute for Environmental Quality. "Asbestos Air Pollution Control" Document 71-8 (1971).
4. Ontario Ministry of the Environment. "Asbestos as a Hazardous Contaminant II (January 1975).

CHAPTER 5

EVALUATION OF ASBESTOS FIBER LEVELS

INTRODUCTION

The main objectives of any air quality evaluation, including asbestos fiber measurements, are:

1. to determine whether an atmosphere is potentially dangerous; and
2. to ascertain whether the dust control measures are efficiently maintaining safe working conditions.

Within the domain of particulate matter in the air, the determination of asbestos fiber levels presents special problems. These problems arise from the fibrous nature and fineness of the particles, their admixture with granular particles and their complex mineralogical makeup. These properties cause problems in the sampling phase as well as in the analytical phase of air quality determinations.

HISTORICAL REVIEW

The early studies of fibrous dust clouds in the U.S. asbestos industries were carried out in the 1930s using techniques employed for granular dusts associated with granite and cement industries.[1] The main instrument employed was the Greenberg-Smith and/or Midget Impinger but others, such as the Konimeter, the Owens Dust Counter and the Thermal Precipitator, which were capable of giving total particle counts, were also used. All these instruments are based on the aspiration of a known volume of air, capture of the particulate phase and subsequent examination and counting under a microscope at a magnification, generally, of 100X. The aspiration volumes varied widely, from 2.5 cc (Konimeter) to 50,000 cc (Midget Impinger).

Given the resolution of the 100X magnification in ordinary light, only particles greater than about 1 μm in size were seen; thus the total count was, at best, a greater than 1-μm nonspecific particle count of general dust levels.

The impinger/optical counting (100X) method was established as a standard method in the U.S. in 1946 and maintained until 1968-1969.[2] Throughout the latter half of the 1960s, serious doubts were expressed about the validity of this method and the accompanying standard, since it was shown that at equal dust levels, large variations in fiber content were found in different industries (5-15% in insulation and 80% in textiles). Also, impinger counts and fiber counts on membrane filters were variously and poorly correlated within and between plants.[3-6] The implication was that only a very small fraction of the fibers were visible in the impinger counting phase and that the larger granular particles were accordingly overemphasized.

During this same period the British developed the method of counting fibers collected on membrane filters (MF). The filter counts were made at 430X magnification using phase-contrast illumination. The British Occupational Hygiene Society (BOHS) established this as a standard method for measuring asbestos fiber levels in 1968. In 1969 the American Conference of Governmental Industrial Hygienists (ACGIH) in the United States proposed that the technique be adopted as the standard method for asbestos fiber measurements. Addingley[7] and Rajhans[3] described and evaluated the efficiencies and the precision/accuracy parameters of the various counting methods.

Ambient air fiber measurements were attempted in 1965 using dispersion staining and light-field microcopy, but no identification or quantification was possible.[8] The use of membrane filters, their dissolution, followed by examination of the dust left behind via an electron microscope at 15-20,000X magnification, was initiated in 1966. Various transmission and scanning electron microscope procedures have been suggested for fiber measurements (fiber number or mass) in ambient air but all these are very recent and little is known of their accuracy or precision. Examples are given by Nicholson *et al.*[9] and Bragg and Taylor.[10]

GENERAL CONSIDERATIONS

The exposure levels to asbestos fiber that may be considered dangerous and the exact mechanism(s) by which health disorders are induced, are questions yet to be resolved and on which extensive research is continuing.

Access to an adequate, simple and rapid method of detecting and quantifying airborne asbestos fiber represents an important input in this research effort as well as a practical necessity for present-day control measures which are being or would be applied in both the occupational and nonoccupational environments.

The principal attributes of a dust cloud, which can or should be evaluated are the following[11]:

1. Total dust concentration: preferably by weight (mg/unit vol);
2. Respirable dust concentration: generally refers to particles less than about 5 μm in aerodynamic diameter; again, a weight measurement;
3. In addition, the respirable fraction can or should be characterized for any of the following significant parameters:
   i. identification of the particles (mineral or otherwise) and their mass and/or number concentration, usually via mineralogical analyses;
   ii. the chemical composition of the total respirable fraction and/or the composition of the individual phases; and
   iii. the morphology of the respirable fraction; total particle size distributions should be characterized and/or relevant size or shape subfractions should be determined; and
4. The length of time during which there is exposure to the above concentrations.

Dust clouds incorporating fibrous particles constitute a complex problem, which Lynch *et al.*[6] described in 1970 as having a number of "degrees of freedom" – in effect, each of the measurable parameters listed above represents one such "degree of freedom."

A survey of the literature pertaining to the evaluation of asbestos fiber levels shows that most authors are often led to conclusions of the following sort:

1. There is, at present, neither a practical absolute method nor any completely satisfactory relative index method for the routine evaluation of asbestos fiber levels in the air.
2. Each of the methods being used or suggested often refer to sample material that is different and, at the analytical level, each measures dust parameters that are different. The resulting correlations between methods are, thus, poor and most often considered to be of qualitative value only.
3. Each method is strongly influenced by one particular parameter of the particulate phase and since the parameters responsible for the biological effects of fibrous dust are only incompletely known, no biologically optimum or absolute method of measurement is effectively possible.

To the extent that there is a consensus concerning any single preferred index of exposure, that provided by the Membrane Filter Technique (see below), as elaborated in 1972 by the National Institute for Occupational Safety and Health (NIOSH), is such a method.[1] There is, however, current interest in a "double" standard based on MF fiber counts and on gravimetic evaluations, to provide for more adequate monitoring of certain occupational environments.[6,8,11]

It follows from this analysis that all past exposure measurements based on nonspecific particle counts (*e.g.*, Impinger, Konimeter, various particle counters; see below) cannot be compared in any significant quantitative fashion with current MF fiber counts.

OCCUPATIONAL EVALUATION METHODS

The evaluation of a dust cloud usually involves two successive operations: (1) sampling, followed by (2) analysis or analyses of the sample. However, certain instruments are direct-reading and measure the particular phase directly in the air without sample collection.

Sampling

Sampling methods may be: (1) discontinuous, short-interval manual sampling (10-30 minutes); (2) continuous long-interval sampling (6-8 hours); or (3) continuous, short-interval automatic monitoring.

The sample collected may be the total dust or the respirable dust fraction, which requires the use of a presampler to effect the size discrimination. There are various types of presamplers (cyclones, impactors, etc.) that have been standardized for granular dust and some that have been modified and standardized for fibrous dust under certain conditions.

In the study and evaluation of complex fibrous dusts, it is recommended that several samples be collected for analysis; further, it is recommended, especially for the continuous and automatic samplers, that the respirable fraction be collected to eliminate the coarse dust, which is a variable of secondary importance. Continuous monitors other than sampling instruments are seldom suitable for asbestos dust.

The length of the sampling interval is defined by the method on which the standards are based. For the MF technique, the period is variable (15-90 minutes) depending on the dust level. However, sampling period must be chosen in accordance with the details of the exposure standard. The time-weighted standard over eight hours can be effectively defined by five or six regular sampling intervals during the eight-hour period. The "maximum allowable level" is usually allowed for an interval of only 15 minutes and would imply nearly continuous monitoring.

Continuous monitoring of fiber levels in key areas of possible high emission sources is considered desirable.[6] The standard MF technique (discontinuous, short-interval and manual) is not practically suited to continuous monitoring. The methods best recognized for this are the continuous short-interval automatic gravimetric techniques; however, the measurement of the fiber content of the resulting samples presents a problem. Measurement procedures have been developed in Germany,[12] but they have yet to be fully disseminated, evaluated and standardized.

Analytical Measurements

A host of direct and indirect analytical techniques have been used for detecting and quantifying asbestos. Harwood and Yamate[13] have compiled and reviewed the various techniques (Table 5.1).

The analytical measurement aspect of evaluating fiber levels is the most problematic. As mentioned, the nature of these fibrous dusts is complex, and the relevant parameters responsible for health disorders are incompletely known. A study of Table 5.1 shows that each method provides data on specific features of the dust cloud. Basically, measurement is needed of the total fiber concentration, data on fiber size distributions, and, particularly, a reliable measurement of the longer respirable fibers (often referred to as being > 5 μm long and < 5 μm in diameter).

The measurement of total fiber concentration is difficult. The normal mineralogical analytical techniques listed in Table 5.1 cannot effectively quantify asbestos minerals and especially chrysotile (see Physical and Chemical Properties of Chrysotile, p. 26), due to interferences and other phenomena. Only tedious visual counts can be converted into acceptable concentration estimates.

Data on fiber size distributions require extensive visual counts using electron microscopes or sophisticated automatic instrumentation (image analyzers) used in conjunction with electron microscopes.

Measurement of the longer respirable fibers (> 5 μ) can be effected by light microscope, as they are in the standard NIOSH membrane fiber technique. However, this measurement is admittedly only partial in nature and most often referred to as being an "index." The problem is one of resolution at 400-430X magnification which is said to be in the range of 0.5 μm (Table 5-1) or 0.1 μm.[14] Electron microscope images show that the total number of fibers > 5 μm is much larger (estimates range from 10-20 times more)[10] than that observable with the light microscope, since many are smaller in diameter than the 0.5-μm resolution limit of the microscope. Low magnification on the electron microscope will, of course, decrease the fiber count using this method.

In summary, both visual counts and mass concentrations are necessary, and the resulting fiber size-mass and number distributions would provide a basis for a complete evaluation of the fiber level.

Table 5.1. Summary of Analytical Methods for Asbestos

Method	Feature Examined	Comment
Light Microscope		
1. With phase contrast at 400X	Morphology	Limit of resolution about 0.5 μm.
2. Dispersion staining	Refractive index and morphology	Skilled operators can distinguish asbestos fibers. Limit of resolution about 0.5 μm.
X-Ray Diffraction	Crystal structure	No information on fiber size or size distribution
Infrared Spectroscopy	Characteristic absorption bands	Ambiguity is possible. No information on fiber size or size distribution.
Atomic Absorption	Elemental composition	No size or size distribution information. High sensitivity for trace elements.
Neutron Activation	Elemental composition	No information on size or size distribution. Specialized nuclear equipment needed.
Emission Spectroscopy	Elemental composition	No size or size distribution information. Both gross and trace constituents determined.
Thermal Analysis TGA and DTA	Weight loss on heating due to dehydroxylation	No information on size or size distribution. Specificity to asbestos not yet resolved.
Scanning Electron Microscope with Microprobe	Surface topology of the fiber and elemental analysis	Most SEMs have a theoretical resolution limit of about 10.0-20.0 nm. Background can give interference.
Transmission Electron Microscope with Microprobe	Shape outline, electron diffraction, and elemental analysis	Resolution limit down to 0.40 nm. Transfer to grid can lead to statistical errors in counting.

Microscope counts can never completely describe a dust, and mass concentrations give no estimate of size distributions. To the extent that characteristic size-mass and size-number distributions can be established for various occupational settings, it may be possible to make a more valid evaluation of the fiber level by such partial counts or by mass concentrations.[15] There is some evidence that fiber concentrations have some of the properties of more ordinary aerosols. Figure 5.1 shows the length distribution of a sampled dust from a fluffing operation. The length distribution is plotted on a probability scale vertically and a logarithmic scale horizontally. A distribution that corresponds to a straight line on this chart is said to be log-normally distributed, since the logarithms of the distribution follow the standard normal or Gaussian distribution. This distribution is frequently found to describe other dusts and aerosols.[16] Figure 5.2 indicates that for this same dust, the diameter of the fiber is also log-normally distributed. This would suggest that a bivariate log-normal distribution could usefully describe this dust. The consequences of this assumption for several dusts have been explored by Bragg;[15] however, the evidence for such a scheme would need extensive documentation before the general properties of a particular distribution could be assigned to industrial asbestos dusts.

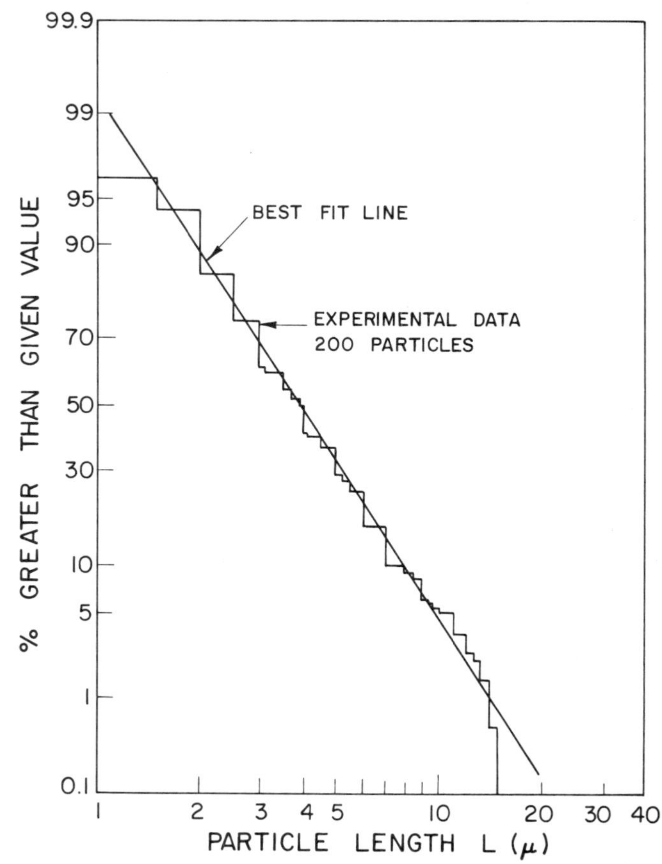

Figure 5.1. Log-normal plot of length of chrysotile fibers.

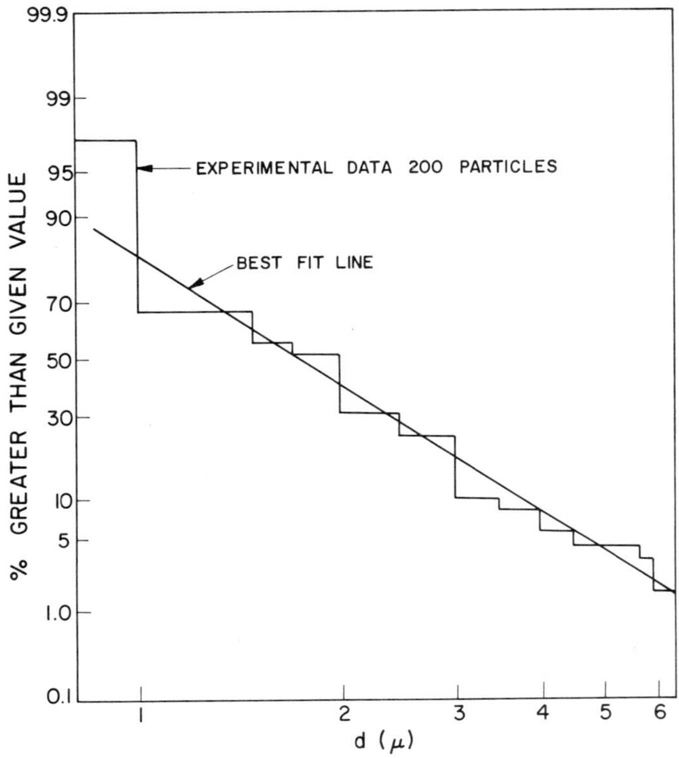

Figure 5.2. Log-normal plot of diameter of chrysotile fibers.

Dust Instrument Types

Measured fiber levels are dependent on the type of dust sampling instruments used and these have been listed. Each of these types has various properties which make them appropriate for different uses.[11]

The direct-reading instruments without sample collection (*e.g.*, Tyndallomoter, Nephelometer, the various particle counters) are rarely recommended as a basis for elaborating a general standard in complex fibrous dusts. The count is generally nonspecific as to particle type and particle size, although some counters can effect a size discrimination that may be applicable to fibrous dusts. The lack of sample is a drawback. These instruments can be used to great advantage, however, for continuous inplant monitoring and control when standardized against accepted evaluation methods.

Of the discontinuous, short-interval manual sampling instruments, those of classical usage such as impingers and konimeters, are considered inapplicable for present-day requirements because they are nonspecific as to size and particle type, limited resolution and other drawbacks. A review of this type of instrument and many others is given in Silverman *et al.*[17]

The MF technique, as elaborated by NIOSH, is the most generally adopted standard method of evaluation (see Chapter 6). It is widely used and has been extensively evaluated. At or near the 2 fibers per cc (f/cc) level (> 5 μm), the reproducibility of the method is said to be ± 35-50% (or 2 fibers ± 1 fiber) and considered sufficiently precise for normal occupational measurements.[18] As noted, however, this measurement is but a partial count and is considered to be an "index" of fiber levels, albeit valid and reproducible. The counts are specific to the respirable fraction, specific as to particle morphology and, by extension, specific to chrysotile and/or the other asbestos minerals. The main drawbacks concern the resolution limit of observation (~0.5 μm), which accounts for the incomplete nature of the count, and the rather tedious and elaborate procedures. In addition, the method cannot be rendered continuous and automatic in any general or obvious manner.

The basic elements of the procedure may be seen from the figures. A small membrane filter sample is taken from the air (Figure 5.3). The sample is mounted on a microscope slide (Figure 5.4) and treated with a solvent to dissolve the filter and render it transparent. The slide is then viewed with a phase-contrast microscope at 400-430 power (Figure 5.5). Figure 5.6 shows a view (without a graticule) of some larger fibers. Better resolution than this can be obtained in practice since the depth of field can be controlled while counting. In performing counting procedures the standard method *must* be used in all details because of the wide variation in results, which is possible with only small changes in procedure. For details of sampling times, fields of view, criteria for counting fibers, solvents, microscope procedures, etc., reference should be made to whichever government document details the procedure for that jurisdiction. At present, most procedures are more or less those of NIOSH.

Instruments of the continuous, long-interval sampling type, such as those of the "simpeds-campeds" family with presamplers, have not, to the authors' knowledge, been used for evaluation of fibrous dusts. Nevertheless, the problems of sample measurement and analysis would apply here as in other cases.

There is, in addition, a family of continuous, short-interval, automatic gravimetric systems with presamplers capable of continuous monitoring. Instruments of this type are the German FH-62A, Vc 25 and MPG II[11,19] and the Quartz Crystal Microbalance. These instruments are reported to give accurate weight measurements of respirable dust over short intervals

Figure 5.3. A closeup of the membrane filter with an asbestos sample on it. A piece is cut for viewing under the microscope.

Figure 5.4. A wedge-shaped section (about $^1/_6$ to $^1/_{10}$ of the total filter area) placed on a microscope slide.

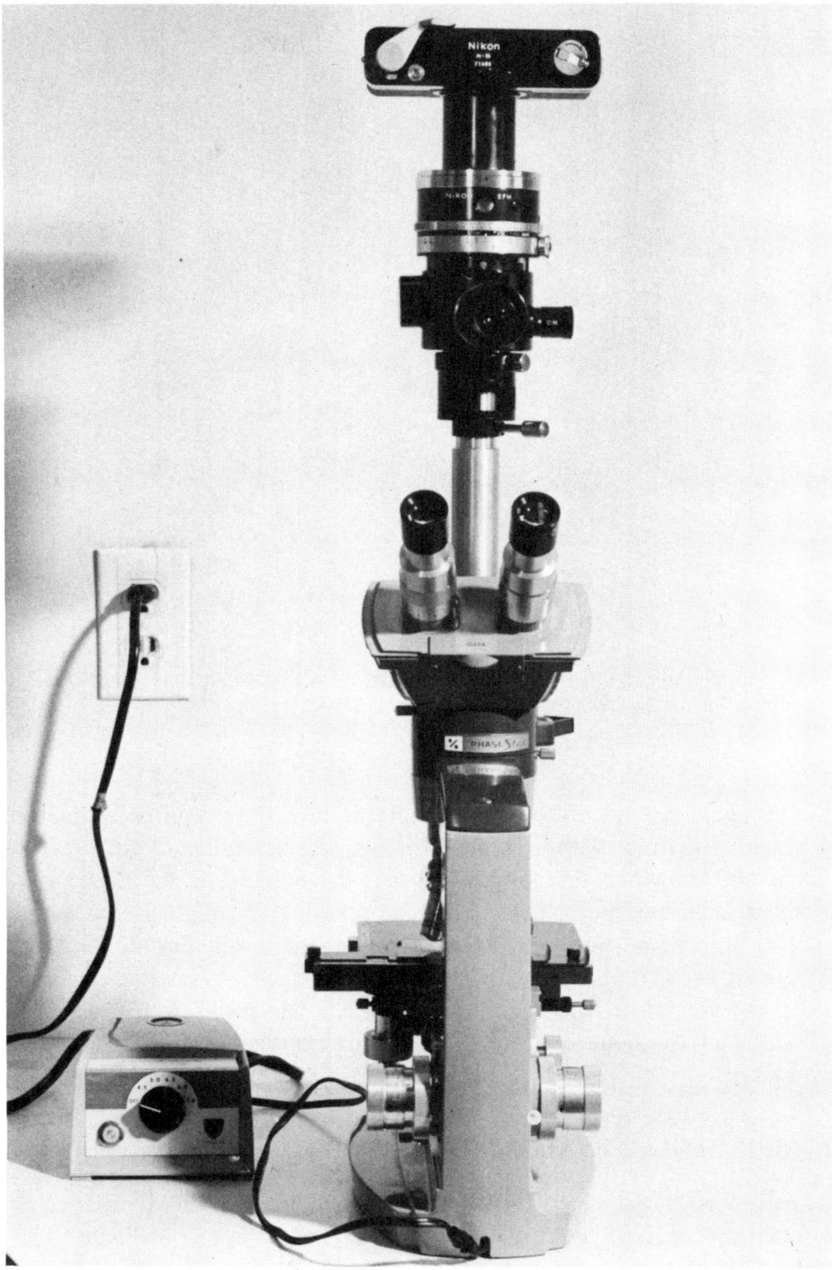

Figure 5.5. Phase constrast microscope with photomicrograph equipment attached.

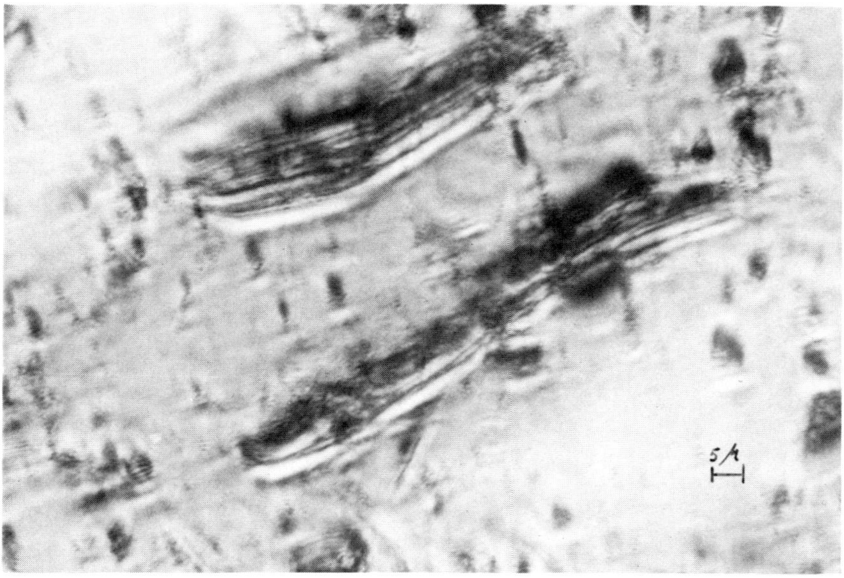

Figure 5.6. Photomicrograph of a group of large fibers seen through the phase contrast microscope. Better focus is possible when viewing. A graticule must be used for fiber sizing.

in continuous fashion. The mass concentration of fiber can be estimated by visual evaluation of the sample, although the precision of this procedure is not known to the authors. It was the opinion of the Beaudry Commission in Quebec[11] that continuous gravimetric sampling could serve a useful purpose in monitoring fiber levels together with the MF technique.

German occupational standards are based on such gravimetric measurements and are set at 0.15 mg/m^3 of respirable chrysotile, which is said to be equivalent to about 3 f/cc $> 5\ \mu$ (MF) as determined in the textile industries.[19]

ENVIRONMENTAL EVALUATION METHODS

Environmental air quality methods have developed from experiences derived from occupational evaluation. As regards asbestos, ambient air differs from occupational air in that there is much less fiber in ambient air, the fibers are much smaller (few or none $> 5\ \mu m$ long) and identification of ambient fiber is more difficult since sources and types of asbestos can be of a mixed type. Thus it was soon shown that standard occupational methods could not be applied to ambient air.

The various existing methods of ambient air evaluation are based on *total* fiber counts at a magnification of 20,000-40,000X using various types of electron microscopes and specialized microprobe identification techniques.

Various sampling and sample preparation techniques are used and the results are given as *total fiber counts/cc* or these are converted into mass measurements per cc. The total count is usually in the million f/cc range whereas the mass measurements are in the tens of ng/m^3.

It is reported[8] that these various approaches to ambient air evaluation are not yet satisfactory, that existing ambient air concentrations are very approximate and that there is little information on the accuracy and precision of the methods. Research in this field is continuing. Here again, the relative importance of counts vs mass measurements arises and only additional fiber size distributions, combined with input from medical research, will help to make a wise choice as to whether the optimal measurement should be a count or a mass concentration.

REFERENCES

1. Health, Education and Welfare, U.S. Department of, Public Health Service, Health Services and Mental Health Administration. "Criteria for a Recommended Standard for Occupational Exposure to Asbestos," National Institute for Occupational Safety and Health, HSM #72-10267, Washington, DC (1973).
2. "Threshold Limit Values of Airborne Contaminants," Adopted at the 30th Annual Meeting of the American Conference of Governmental Industrial Hygienists, St. Louis, MO, May 13, 1968.
3. Rajhans, G. S. "Fibrous Dust–Its Measurement and Control," *Can. Inst. Mining Metall. Bull.* 63:900-910 (1970).
4. Ayer, H. E., J. R. Lynch and J. H. Fanney. "A comparison of Impinger and Membrane Filter Techniques for Evaluating Air Samples in Asbestos Plants," *Annals N. Y. Acad. Sci.* 132, Part. 1:274-287, December 1965.
5. Leathart, G. I., and J. T. Sanderson. "Some Observations on Asbestosis," *Ann. Occup. Hyg.* 6:65-74 (1963).
6. Lynch, J. R., H. E. Ayer and D. C. Johnson. "The Interrelationships of Selected Asbestos Exposure Indices," *Am. Ind. Hyg. Assoc. J.* 31:598-604 (1970).
7. Addingley, C. G. "Asbestos Dust and its Measurement," *Ann. Occ. Hyg.* 9:2 (1966).
8. Ontario Ministry of the Environment. "Asbestos as a Hazardous Contaminant II" (January 1975).
9. Nicholson, W. J., A. N. Rohl and I. Weisman. "Asbestos Contamination of the Air in Public Buildings," EPA-450/3-76-004, National Technical Information Service #PB-250 980 (October 1975).

10. Bragg, G. M., and J. D. Taylor. "Asbestor Fiber Enumeration Using the Scanning Electron Microscope," Contract Report 1121-3 to Ontario Ministry of Health (1974).
11. Comité d'Étude sur la Salubrité dans l'Industrie de l'Amiante, Québec Service de la reprographie, Québec (April 1976), pp. 282-285.
12. Trudeau, P., and G. Grassmück. Personal Communication (1976).
13. Harwood, C., and G. Yamate. "The Detection and Quantification of Asbestos Present in the Environment," *3rd Int. Cong. Physics Chem. Asbestos,* Quebec (1975).
14. Kennan & Lynch. "Techniques for the Determination, Identification and Analysis of Fibers" *Am. Ind. Hyg. Assoc. J.,* September-October 1970, p. 596.
15. Bragg, G. M. "A Two Parameter Model of Fibrous Dusts" Contract Report 1121-2 to Ontario Ministry of Health (1973).
16. Cadle, R. D. *Particle Size,* 33 (New York: Reinhold Publishing Corp., 1965).
17. Silverman, L., C. E. Billings and M. W. First. *Particle Size Analysis in Industrial Hygiene,* (New York: Academic Press, 1971).
18. Rajhans, G. S., and G. M. Bragg. "A Statistical Analysis of Asbestos Fiber Counting in the Laboratory and Industrial Environment," *Am. Ind. Hyg. Assoc. J.* 36:909-915 (1975).
19. Schutz, A. "Technical Standard Values for the Admissible Working Place Concentration of Chrysotile Asbestos," *Staub Reinhaltung* (1973), p. 33.

CHAPTER 6

DEVELOPMENT OF ASBESTOS STANDARDS

INTRODUCTION

There has always been an attempt on the part of scientists to establish a hygienic standard for asbestos that is biologically most appropriate in limiting the hazard. However, since asbestos was established as a human carcinogen, it has been difficult to establish a no-risk level either for occupational or nonoccupational exposure. There is always some small risk to health as long as there is any airborne dust in the environment. Nevertheless, an allowable exposure level must be set and defined by an index, and exposure based on the theory that exposure up to certain levels can be tolerated without incurring undue risks.

OCCUPATIONAL STANDARDS

Development of Occupational Standards

The occupational or work-environment level is normally referred to as a Threshold Limit Value (TLV) or Time-Weighted Average (TWA). These values refer to time-weighted concentrations for a 7- or 8-hour work day and 40-hour work week. They refer to airborne concentrations to which it is believed that nearly all workers may be repeatedly exposed day after day without adverse effect. In 1946 the American Conference of Governmental Industrial Hygienists proposed that the TLV for asbestos dust be set at 5 million particles per cubic foot (mppcf) as determined from impinger samples. This value was recommended by Dressen *et al.* after a study of 541 employees in three asbestos textile plants using chrysotile asbestos.[1]

In 1960 an occupational standard for asbestos was published in Britain by the Ministry of Labor. This was taken from the annual list of TLVs published by the American Conference of Government Industrial Hygienists

(ACGIH). In 1965 the British Occupational Hygiene Society (BOHS) formed a Hygiene Standard Committee.[2] This committee, after critical examination of the ACGIH standards, recommended the use of the ACGIH standard for asbestos until they were able to establish their own standard.

In 1967 Cooper[3] observed that the TLV of 5 mppcf for asbestos rested on shakier evidence than most TLVs. His chief argument was that a large proportion of asbestos fibers have diameters below the resolving power of the light microscope used for impinger counts and are, therefore, not counted at all. He further reported that asbestosis appeared in insulation workers who were exposed to concentrations much below the TWA of 5 mppcf. Leathart and Sanderson[4] also considered the TLV of 5 mppcf to be too high. In January 1968, the ACGIH issued the following statement: "A limit to 5 mppcf, based on impinger samples counted by light-field techniques, is satisfactory to control exposures to most forms of asbestos. Crocidolite, however, has been shown to produce mesotheliomas in addition to the common asbestotic inflammations. Since no safe limit can be established for this form of asbestos at this time, until more definitive data are obtained, it is recommended that workers exposed to crocidolite be equipped with air-supplied helmets."[5] However, by this time, the Hygiene Standards Committee of the BOHS under the chairmanship of Professor Lane had already completed an environmental and medical study at an asbestos textile factory in England, in which investigations were made of 290 of the men employed in the period 1933-1966.[6] The asbestos dust concentrations to which different workers had been exposed varied from 1-27 f/cc measured by the MF technique in which the number of fibers greater than 5 μm in length were counted. The problem of relating prevalence of asbestosis with concentration and with duration of exposure was resolved by assuming that the cumulative amount of asbestos remaining in the lungs from a given exposure is equally dependent on the dust concentration and on the duration of exposure to that concentration. Thus, the exposure of an individual was expressed as f-yr/cc, that is, the product of the average concentration during working hours and and his years of exposure. Exposure expressed in this way was compared to the prevalence of asbestosis, and a curve was developed[7] (Figure 6.1).

One criticism of this curve was the use of a continuous scale starting at zero. Doses that give zero response are, in fact, never known with absolute certainty, since the sensitivity of measuring instruments is limited. However, the main conclusion from the curve was that for an accumulated exposure of 100 f-yr/cc, it is probable that the risk of being affected, to the extent of having the earliest clinically demonstrated signs, is less than 1%. That is, 2 f/cc for 50 years, 4 f/cc for 25 years or 10 f/cc for 10 years.

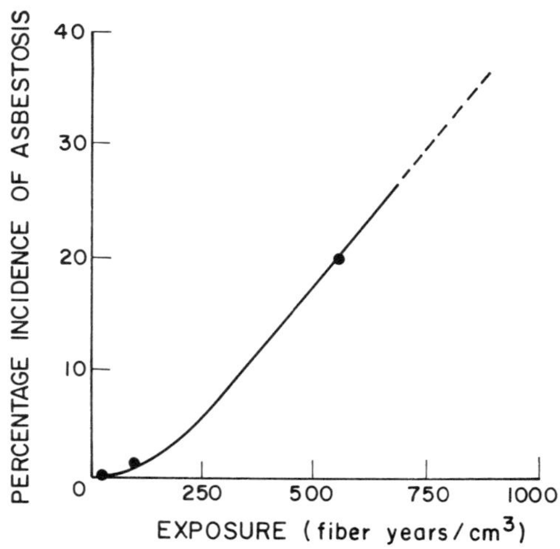

Figure 6.1. Response to chrysotile asbestos exposure.

American Standards

In 1969 the ACGIH proposed a TLV of 12 fibers >5 μm in length per cc by the MF technique or 2 mppcf by the impinger technique. This was probably based on the work of Lynch *et al.*[8] in asbestos textile plants where 1 mppcf (impinger) was shown to be equal to 5.9 f/cc. In 1970 the ACGIH proposed a lower asbestos dust limit of 5 f/cc by the MF technique and eliminated the impinger standard. However, this TLV remained on the intended changes list until 1973 when it was first adopted as the TLV and continues to be the one today. This TLV, if continued, would result in an accumulated exposure of 250 f-yr/cc over 50 years and an extrapolation of the curve shown above would indicate that this would result in more than 2% risk of contracting asbestosis.

In 1972 NIOSH recommended the following criteria for asbestos dust[9]: "Occupational exposure to airborne asbestos dust shall be controlled so that no worker shall be exposed to more than 2.0 asbestos fibers per cubic centimeter (cc) of air based on a count of fibers greater than 5 micrometers (>5 μm) in length determined by the membrane filter method at 400-450X magnification (4 mm objective) phase contrast illumination, as described (here), determined as a time-weighted average (TWA) exposure for an 8-hour work day, and no peak concentration of asbestos to which workers are exposed shall exceed 10.0 fibers/cc >5 μm as determined by a minimum sampling time of fifteen minutes."

Along with the above recommendation the NIOSH also issued several other statements:

1. The standard is amenable to techniques that are valid, reproducible and available to industry and government agencies. It will be subject to review and will be revised as necessary.
2. The recommended standard is designed primarily to prevent asbestosis. For other diseases associated with asbestos, there is insufficient information to establish a standard to prevent such diseases including asbestos-induced neoplasms by any all-inclusive limit other than one of zero. Nevertheless, a safety factor has been included in arriving at the concentration level that will reduce the total body burden and should more adequately guard against neoplasms.

A ceiling value of 10.0 f/cc that was not to be exceeded was included to reduce the possibility of the short-term heavy exposures to asbestos that have been reported to cause mesothelioma. In addition, this should reduce the likelihood of diseases (malignant and nonmalignant) resulting from exposure in excess of 30 years with very long latent periods. The U.S. Government published the following in the *Federal Register*[10]: "The 8-hour time-weighted average airborne concentration of asbestos dust to which employees are exposed shall not exceed 5 fibers per milliliter greater than 5 microns in length, as determined by the membrane filter method at 400-450X magnification (4 millimeter objective) phase contrast illumination. Concentrations above 5 fibers per milliliter but, not to exceed 10 fibers per milliliter, may be permitted up to a total of 15 minutes in an hour for up to 5 hours in an 8-hour day." On June 7, 1972 (see *Federal Register* for October 18, 1972[11]), OSHA promulgated a standard which limits the worker exposure to an 8-hour TWA of 5 f/cc of air, and the ceiling limit of 10 f/cc. They also mentioned that effective July 1, 1976, the 1972 rule would reduce the exposure limit to an 8-hour exposure limit of 2 f/cc. Selikoff *et al.*[12] criticized the standard on the basis that the fibers are made of fibrils 300-350 Å in diameter which cannot be seen with the optical microscope, being far below the limits of its resolution.

On October 9, 1975, the *Federal Register*[13] proposed a standard of 0.5 f/cc for an 8-hour TWA (or 500,000 f/m^3) and likewise a reduced ceiling of 5 f/cc (or 5 million f/m^3) for any period not exceeding 15 minutes. In addition, they also proposed the following:

1. modification of definitions as follows: "asbestos" includes chrysotile, amosite, crocidolite, tremolite, anthophyllite, actinolite and every product containing any of these minerals; and "asbestos fiber" means a particulate form of asbestos, longer than 5 μm, with a length-to-diameter ratio of at least three-to-one and with a maximum diameter of 5 μm;

2. establishment of requirements for engineering controls, work practices and supplemental use of respirators and protective clothing;
3. revision of procedures for monitoring;
4. work practices be followed for certain operations and processes involving asbestos;
5. revision and update of the requirements for respirators, referencing NIOSH as the approving authority; the proposal would not render any existing respirators obsolete, but would enable a greater variety of respirators to be used under different conditions of exposure;
6. establishment of additional requirements for regulated areas, hygiene facilities and practices (including a requirement for change rooms and showers), worker training and information programs, and warning signs and labels;
7. specific medical surveillance and minimum data on medical and monitoring records;
8. job reassignment in case of medical unsuitability; extension of the period during which the employer must retain monitoring and medical records to 40 years, or employment plus 20 years, whichever is longer; and
9. provision of procedures for the transfer of employees' medical and monitoring records when one employer succeeds another, or when an employer goes out of business.

The proposed standard did not apply to the construction industry, and separate rule-making to revise the asbestos standard for construction industry was planned. The Insulation Hygiene Progress Report[14] critically reviewed the proposed OSHA standard and issued the following statement: "This recognition that the construction industry is unique is highly desirable. Earlier data on the effectiveness of monitoring in the construction industry (see *Insulation Hygiene Progress Reports,* Summer 1973), have shown that few dust counts are ever taken in this segment of the industry. Thus, specification of procedures is of much greater importance here. Individual workers must be able to judge, on the basis of materials and methods rather than dust counts, whether they are working in a safe environment.

As much of insulation is now asbestos-free, work practices for installation of new material can be developed that would lead to a TWA less than that proposed in the new standard. With care, even rip-out can be done safely, but special precautions must be taken. Wetting techniques have now been developed that would allow removal of asbestos to be accomplished within the proposed regulations."

While proposing the above, OSHA also mentioned the major issues concerning the standard:

"(1) Whether, and by what date, the permissible limit for an 8-hour time-weighted average exposure to asbestos should be lowered to

500,000 fibers per cubic meter (0.5 fiber per cubic centimeter), and, if so, whether such an exposure limit would provide an adequate margin of safety to protect employees against known and suspected workplace hazards of asbestos.

(2) Whether, and by what date, the permissible ceiling limit should be reduced to 5 million fibers per cubic meter (5 fibers per cubic centimeter), and, if so, whether such an exposure limit would provide an adequate margin of safety to protect employees against known and suspected workplace hazards.

(3) Whether the proposed change in the definitions of "asbestos" and "asbestos fiber" would clarify the standard's intended scope, and properly relate to known or suspected workplace hazards.

(4) Whether the proposed sampling period for the determination of ceiling concentrations would be appropriate and adequate.

(5) Whether the proposed procedures for initial and subsequent exposure monitoring would be appropriate and adequate.

(6) Whether the proposed retention period for medical and monitoring records would be appropriate and adequate.

(7) Whether the requirement for specific data on medical and monitoring records would be appropriate and adequate.

(8) Whether the procedure for transfering the medical and monitoring records of former employers would be appropriate and adequate.

(9) Whether the information gathering requirements of the proposal would create any undue administrative and economic burdens on employers, particularly for those employing small numbers of employees, or workforces which are highly transient in nature.

(10) Whether the revised schedule for allowable respirator use would be appropriate and adequate for protecting workers against exposure to asbestos.

(11) Whether the provision for job reassignment on account of medical unsuitability is appropriate, and, if so, whether it effectively furthers the purposes of the Act.

(12) Whether the appendices are appropriate and otherwise in the best interests of worker health and safety.

(13) To what extent, if any, should the proposal be modified to reflect a concern for workplaces which are of a nonfixed nature or otherwise engage a highly transient workforce.

(14) Whether complicance with the proposal as a whole would be technologically and economically feasible for all affected industries; and particularly for employers engaging small numbers of employees.

(15) Whether the standard should have a delayed effective date for any industry sector and, if so, the extent to which a phased schedule for compliance would be appropriate.
(16) What are the environmental and inflationary impacts of the proposal."

OSHA invited comments on the above issues including the appropriateness and feasibility from the asbestos industry. On February 4, 1976,[15] some 65 representatives from companies and trade associations representing manufacturers and processers of asbestos products in the United States participated in a meeting held in Washington, DC and overwhelmingly endorsed that the "2-fiber" level has been or can be attained by application of engineering technology but the proposed "0.5-fiber" level is unnecessary, impracticable and lacks medical justification.

The situation was further altered in late 1976. The following extract from *Asbestos* illustrates the situation as of January 1977[16]: "By letters dated December 15, 1976, to the U. S. Occupational Safety and Health Administration (OSHA) and the U. S. Department of Health, Education and Welfare (HEW), Dr. John F. Finklea, Director of the National Institute for Occupational Safety and Health (NIOSH), transmitted the NIOSH study on the "Re-examination and Up-date of Information on the Health Effects of Occupational Exposure to Asbestos." The study reviewed the Institute's evaluation of the health effects of exposure to asbestos and recommended that OSHA reduce the Standard for Occupational Exposure to Asbestos." The study reviewed the Institute's evaluation of the health effects of exposure to asbestos and recommended that OSHA reduce the Standard for Occupational Exposure to Asbestos to 0.1 f/cc (100,000 f/m^3). Dr. Finklea said the previous recommendation by NIOSH of 2-fiber standard was designed primarily to prevent asbestosis, since insufficient information then existed to establish a standard to prevent other asbestos-related diseases such as pulmonary, pleural and peritoneal neoplasms. NIOSH says the 0.1 recommendation is based on lowest concentration at which asbestos fibers can be monitored reliably using phase contract microscopy. The study addresses only the health effects of exposure to asbestos and the proposed environmental limit, but NIOSH intends to address other provisions of OSHA's proposed revision to the Asbestos Standard as soon as possible.

In its preliminary statement regarding the NIOSH document, the Asbestos Information Association/North America (AIA/NA) commented that studies cited in the document do not define the levels of airborne asbestos dust to which individuals were exposed. The levels of exposure were most probably much higher than those now being experienced in the workplace. The current OSHA Asbestos Manufacturing Standard sets a limit of 2 f/cc of air longer than 5 μ in an 8-hour TWA with a ceiling limitation against any exposure in excess of 10 f/cc. AIA/NA says there is no evidence to support the proposition that the present 2-fiber limitation is not safe for exposed populations. Further, contrary to the NIOSH statement, the state-of-the-art and practical considerations rule out sampling and analytical functions at a 0.1 f/cc level by phase contrast microscopy at 400-450X magnification. In most products in which asbestos is a vital ingredient, the fibers are bonded or locked in, and the risk of exposure to respirable particles of asbestos from normal use of these products is negligible.

British Standards

In 1973 the Hygiene Standard Committee (UK) reviewed the 1968 BOHS standard for chrysotile and recommended that no changes be made at the present time, but the standard be kept under review.[17] They also reviewed the information on the results of human exposure to airborne amosite asbestos with animal experiments. They recommended that the standards for amosite be the same as those for chrysotile.

The BOHS has recommended a level of 0.2 f/cc for crocidolite.[18] The BOHS also gives the following alternative indices for chrysotile asbestos. Respirable mass:

0.03 mg magnesium/m^3
0.04 mg SiO_2/m^3
0.10 mg ash/m^3
0.12 mg asbestos/m^3

Thermal Precipitator:	25 ppcc greater than 1 μm after incineration
Impinger:	10 ppcc
Royco Particle Counter:	2 ppcc greater than 4 μm

It should, however, be realized that the above standards refer to conditions in textile factories using chrysotile asbestos and the equivalents may not be applicable in other places.

It was further recommended that the following table be used to designate certain ranges of dustiness:

Category	Dust Concentration (f/cc)
Negligible	0-0.4
Low	0.5-1.9
Medium	2-10.0
High	$>$ 10.0

Canadian Standards

In Canada, there is no uniform standard. Each province has its own asbestos standards. An attempt is made here to summarize the provincial standards.

Quebec. The standard to be applied is 5 f/cc,[19] but the Beaudry Commission is looking into the status of the regulation, and there is possibility of a reduction to 2 f/cc or the equivalent.

Ontario. Ontario followed the ACGIH recommendation until 1972 and then lowered the standard to 2 f/cc, mainly because: (1) the UK has been using this standard since 1971; (2) the NIOSH criteria document recommended this limit; and (3) the toxicities of various types of asbestos are not well understood and, hence, a built-in safety factor was needed. For crocidolite the standard is 0.2 f/cc.[20] There is also a ceiling value of 10 f/cc for a 15-minute exposure for chrysotile and 1 f/cc for crocidolite.

Alberta. On April 15, 1975, Alberta promulgated the following occupational health standards:[21]

1. For asbestos dust or fiber, all forms, the time-weighted average airborne concentration shall not exceed 2 f/cc greater than 5 μm in length of air, as determined by the membrane filter method of evaluation at 400-450X magnification using phase contrast illumination.
2. For asbestos dust or fiber, all forms, the maximum (ceiling) airborne concentration shall not exceed 10 f/cc greater than 5 μm in length of air as determined by the membrane filter method of evaluation at 400-500X magnification using phase contrast illumination.

British Columbia. The BC government has long had a standard of 5 mppcf (impinger). However, it has revised its standards and has mentioned a TWA of 2 f/cc greater than 5 μm in length for chrysotile and 0.2 f/cc for crocidolite in Industrial Health and Safety Regulations (February 1976). It has also mentioned a ceiling of 5 f/cc for 15-minute exposure. This is half the ceiling level of 10 f/cc recommended by Ontario.

Saskatchewan. The province has a TLV of 2 f/cc for all asbestos varieties.

Nova Scotia. At present, the standard is 5 f/cc, but new facilities will be required to meet the standard of 2 f/cc.

The Working Group on Asbestosis (Canada) recognized the documented and suspected health hazards associated with asbestos exposures and recommended a uniform early acceptance of a maximum tolerable level of 2 f/ml, time-weighted average.[23] The practical objective, however, should be to reduce asbestos dust in the workplace to the lowest possible level.

Feasibility of the 2 f/cc Standard

The joint ACGIH/AIHA Aerosol Hazards Evaluation Committee,[18] after reviewing various standards including the mass concentration, concluded that "at present the only practical way to determine fiber concentration is by counting fibers which have been mounted on a suitable substrate so that they can be identified as individual fibers." Based on this conclusion, we may assume that for the next few years, at least, government standards will be based on some level of fiber count.

Turning now to the particular value of 2 f/cc, we see that this is rapidly becoming the standardized value for chrysotile in Britain, the United States and Canada. It is worthwhile to analyze the feasibility of the standard. It should be realized that 2 f/cc is a TWA over an 8-hour period. In other words, some excursion during the period can be permitted, but the excursion is limited to 10 f/cc in most jurisdictions. Now, if one supposes that a worker is exposed to 10 f/cc for 15 minutes he could be exposed to 1.7 f/cc for the rest of his shift, *i.e.* (15 min x 10) + (465 min x 1.7) = 480 x 2.

Lemen *et al.*[24] showed that more than 50% of the operations in textile operations were able to meet the 2 f/cc limit and control technology to meet the 2 f/cc level existed in all operations in the friction product industry. Rajhans *et al.*[25] have surveyed several plants in Ontario which have been developing their systems to meet a 2 f/cc standard. Figure 6.2 shows the results of measurements over a period of four years in a brake lining and clutch facing plant. The open bars indicate the average values of the samples taken at a given time and the shaded bars indicate the range of the samples taken. It can be seen that this particular plant is now operating well within the standard. Figure 6.3 shows similar results in an asbestos piping and paneling plant. In this case, while the average is well below 2 f/cc, there are frequent dust concentrations above this value. In spite of this, there would seem to be evidence that this plant is also capable of meeting the standard. Extensive data[25] have been collected in Ontario, and there is continuing evidence of the ability of most plant operations to meet this standard.

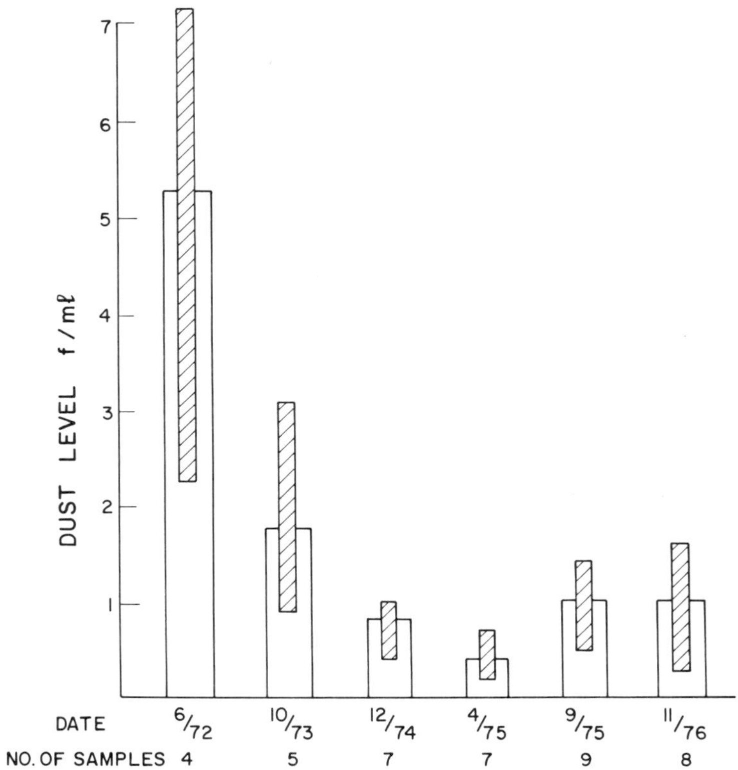

Figure 6.2. Measured dust levels in a brake lining and clutch facing plant. Open bar records average reading and hatched bar records extreme values recorded.

ENVIRONMENTAL STANDARDS

In 1971 the U. S. Environmental Protection Agency proposed tentative emission standards concerning the following:[9]

1. Emissions were not to exceed those which would be emitted from operations if proper engineering control had been installed (*i.e.*, fabric filter, cyclone gas cleaning devices).
2. There should be no visible emissions of particulate.
3. Open spraying of asbestos should be stopped.
4. Use of asbestos for surfacing or resurfacing of roads should be stopped.

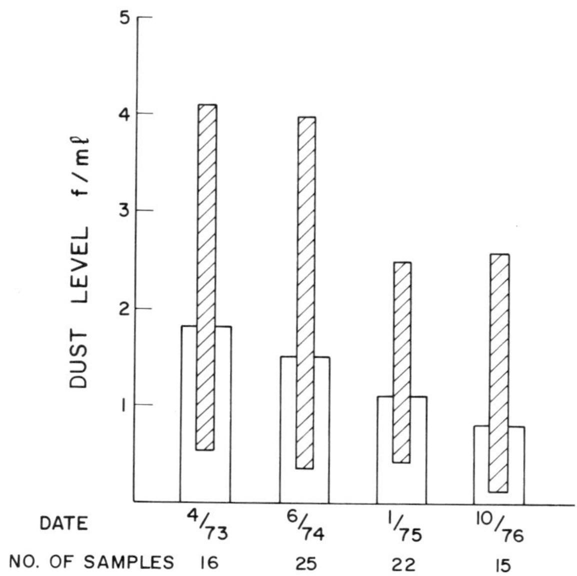

Figure 6.3. Measured dust levels in an asbestos piping and panelling plant. Open bar records average reading and hatched bar records extreme values recorded.

The use of procedural standards and visible emissions as the basis for evaluation for compliance with the standard are designed to minimize emission to the atmosphere. At about the same time the U. S., through the Air and Industrial Hygiene Laboratory, California State Department of Health, recommended that the total fiber count downwind of an asbestos source not exceed 10 times the total fiber count upwind of that source. The Department of Environmental Protection, State of Connecticut, and the U. S. Environmental Protection Agency had the problem under study, and required only that there be no visible emission or certain specified control equipment must be used, *e.g.*, a baghouse, high-energy scrubber, etc.

Ontario has an emission guideline,[26] which is 5 $\mu g/m^3$ averaged over a half hour at point of impingement. This guideline refers to the maximum concentration of asbestos averaged over a half hour resulting from the plume from any source, stack or vent, impinging on any receptor, a school, shopping center, ground, etc., at the time when the wind is blowing from the source directly toward the receptor.

Since wind direction is variable and since more than one source may impinge on a given receptor at one time, the emission guideline is developed to encompass these variables in such a way that, if every source meets the emission guideline, the ambient air quality guideline will not be exceeded. Thus, there is no direct numerical relationship between the emission guideline and the ambient air quality guideline. The ambient air quality guideline is the maximum concentration, averaged over 24 hours, to be found anywhere in the community.

Table 6.1 summarizes some of the air quality standards in force today:[26]

Regulating Agency	Asbestos-Concentration Standard
California	
Air and Industrial Hygiene Laboratory, California State Department of Health	Total fiber count downwind of an asbestos emission source must not exceed 10 times the total fiber count upwind of that source
Connecticut	
Department of Environmental Protection, State of Connecticut	30 ng/m^3 (based on 30-day average sample)
Ontario	
Ontario Ministry of the Environment, Air Resources Branch	Interim guideline of 5 $\mu g/m^3$

The Final National Emission Standards for Hazardous Air Pollutants–Asbestos and Mercury, promulgated by the U. S. Environmental Protection Agency, were published in the *Federal Register* of October 14, 1975,[27] and were effective as of that date. The following are some of the highlights of this new document as it concerns asbestos:

1. The "no visible emissions" cluse for the waste disposal of asbestos materials remains in effect.
2. Asbestos mills are exempt from the requirement to wet tailings when disposal site temperatures are below 15°F,
3. Only asphalt concrete manufacturing plants that use commercial asbestos are affected.
4. Local exhaust ventilation systems may be used during certain renovation operations.
5. Containers of asbestos waste may display either an EPA or an OSHA warning label.
6. Clarification is given of the definitions for "planned renovation" and "emergency renovation" operations.
7. Wastes generated by operations that use sprayed-on materials containing less than 1% of asbestos by weight to insulate or fireproof buildings, structures, pipes and conduits are not covered by the new rules.

REFERENCES

1. Dressen, W. C., J. M. Dallavalle, T. I. Edwards, J. W. Miller, R. R. Sayers, H. F. Easom, and M. F. Trice. "A Study of Asbestosis in the Asbestos Textile Industry," *Publ. Health Bull.* No. 241, Washington, DC (1938).
2. Committee of Hygiene Standards of the British Occupational Hygiene Society. "Hygiene Standards for Chrysotile Asbestos Dust," *Ann. Occup. Hyg.* 11:47-69 (1968).
3. Cooper, W. D. "Asbestos as a Hazard to Health," *Arch. Environ. Health* (September 1967), pp. 285-289.
4. Leathart, G. I., and J. T. Sanderson. "Some Observations on Asbestosis," *Ann. Occup. Hyg.* 6:65-74 (1963).
5. "Threshold Limit Values of Airborne Contaminants." Adopted at the 30th Annual Meeting of the American Conference of Governmental Industrial Hygienists, St. Louis, MO, May 13 1968.
6. "Hygiene Standards for Chrysotile Asbestos Dust," *Ann. Occup. Hyg.* 11 (1968).
7. Roach, S. A. "Hygiene Standards for Asbestos," *Ann. Occup. Hyg.* 13:7-15 (1970).
8. Lynch, J. R., H. E. Ayer, and D. C. Johnson. "The Inter-relationships of Selected Asbestos Exposure Indices," *Am. Ind. Hyg. Assoc. J.* 31:598-604 (1970).
9. Health, Education, and Welfare, U. S. Department of, Public Health Service, Health Services and Mental Health Administration. "Criteria for a Recommended Standard for Occupational Exposure to Asbestos," National Institute for Occupational Safety and Health, HSM #72-10267, Washington DC (1973).
10. *Federal Register* 36(234):2; 342-2; 343 (December 7, 1971).
11. *Federal Register* 37(22):142-22; 144 (October 18, 1972).
12. Selikoff, I. J., E. C. Hammond and H. Seidman. "Cancer Risk of Insulation Workers in the United States," in *Biological Effects of Asbestos,* International Agency for Research on Cancer, Lyon, 1973, pp. 209-216.
13. *Federal Register* 40(197) (October 9, 1975).
14. "New Asbestos Standard Proposed," *Insulation Hyg. Prog. Rep.* 7 (3/4) (1975).
15. *Asbestos* (February 1976), pp. 8-90.
16. *Asbestos* (January 1977), p. 34.
17. "Review of The Hygiene Standard for Chrysotile Asbestos Dust," *Ann. Occup. Hyg.,* 16:7 (1973).
18. ACGIH-AIHA Aerosol Hazards Evaluation Committee. "Background Documentation on Evaluation of Occupational Exposure to Airborne Asbestos," *Am. Ind. Hyg. Assoc. J.* 36:91 (1974).
19. Comite d'etude sur la Salubrite dans l'Industrie de l'Amiante (April 1976), pp. 282-285.
20. Occupational Health Protection Branch, Ontario Ministry of Health. "Asbestos Data Sheet," (1st draft), (August 1976).
21. Verma, D. K., Government of Alberta. Personal Communication (1975).

22. Government of British Columbia. "Industrial Health and Safety Regulation, Workers Compensation Board (Second Draft Amendments)," Victoria, BC (February 1976).
23. Health and Welfare Canada. "Report of the Asbestos Working Group," Queen's Printer, Ottawa (February 15, 1976).
24. Lemen, R. A., J. M. Dement and R. E. Seiter. "New Asbestos Standard–Feasibility of Compliance Within Selected Asbestos Industries," NIOSH Internal Report (1972).
25. Rajhans, G. S., G. M. Bragg and J. S. Morton. "A Review of Asbestos Exposures in Ontario, *Am. Ind. Hyg. Assoc. J.* (In press).
26. Ontario Ministry of the Environment. "Asbestos as a Hazardous Contaminant II" (January 1975).
27. *Federal Register* (October 14, 1975).

CHAPTER 7

OCCUPATIONAL EXPOSURE TO ASBESTOS

INTRODUCTION

Although asbestos has been known since ancient times, its industrial uses and health effects date only from the early part of this century. Asbestos has also been found in community air samples and in city dwellings. Whereas the mortality studies have been widely used to measure the effect of asbestos exposure in working populations (occupational), the association between community air pollution (nonoccupational) exposures and increased lung diseases has not been established. One of the reasons has been the lack of a method for the quantitative assay of airborne asbestos fibers which are present in much lower concentrations than other airborne particles in ambient air.

The method for occupational exposure evaluation has traditionally been visual counts of asbestos dust particles. However, almost every type of dust sampling instruments has been used for sampling asbestos dust clouds.[1] In the United States and Canada, impingers and konimeters have been in use since 1930, but due to various methods of recording observations, the results were difficult to compare. It is hoped that the current standardized membrane filter method of fiber counting will help collate the sampling data from various parts of the world. An attempt is made in this and the following chapter to present some examples of occupational and nonoccupational exposures.

In each case, we will discuss the emission mechanism, the evidence for exposure in that industry, where available, and briefly outline the types of control now typically employed. Detailed discussions of the control mechanisms will follow in Chapter 10.

MINING AND MILLING

Emissions

It has long been realized that asbestos mills are dustier than mines. There are several areas in the process outlined in Figure 2.2 which can cause excessive exposures. In particular, the exposure in mills in the crushing and "fiberizing" processes can be high; nevertheless the mining process can also be a source of both occupational and environmental contamination.

In the modern mine there are a number of operations causing dust emissions. Many of these are common to both underground and open pit operations. Drilling is a dust-creating process that must be controlled. For air-flushed drills, collection by ventilation procedures is often used. Wet drilling methods may not be possible in freezing conditions except when it is possible to heat the flushing water to prevent freezing.[2] Ice in open pit operations can of course be a safety hazard. Also mitigating against wet drilling is the potential for drill cuttings to cement themselves into the hole if the water concentration is low. Air flow rates are high in the larger dry drills and large amounts of contaminated air can be released.

Blasting is an obvious dust creator also. The difficulty in reducing dust arises from the very fast creation of new surface area and the very large gas velocities past these areas. Good blasting practice requires, however, that gases created in a blast be used to fracture rock and that as little as possible of the gases escape directly to the atmosphere. Surface spraying has been considered[3] but the large surface areas suggest that the chance for control in this way is minimal. Liquids and pastes have also been attempted with some success.[2] Good blasting practice minimizes the need for additional or secondary blasting and this also is consistent with antipollution requirements.

Scraping, shoveling and trucking of ore all cause some dusting as do the various operations related to removal of overburden. There is little control for most of these operations, although trucks can be covered with plastic or tarpaulins. Road dust is a problem in summer, and constant wetting is required in dry weather. Road dust is not a problem in winter. Wetting is again a partial solution at best because, in the milling process, the water must be removed.

In open pit operations, wind erosion of exposed ore can cause extensive spreading of fiber. Economical methods of control are not presently available, and this type of mechanism can be expected to be a major source of any fiber in the region of open pit operations or large uncovered storage areas. In underground operations, care must be taken to prevent the mine ventilation exhaust from being a source of pollutant.

Milling represents a much more difficult problem for occupational exposures. For example, bagging of fiber can be a very dusty process depending on the method of bagging. The recent introduction of pressure-packing methods and leakproof bags has reduced the exposure to a much lower degree than it was when hessian bags were used. The hessian bags presented a dust hazard not only to the baggers but to transport workers and dockers moving asbestos cargoes in ships' holds and warehouses.

In the ore preparation stage (Figure 2.2), conveyors feed a jaw crusher producing maximum dimensions of the output of around 6 inches. Hand sorting is frequently used to separate large pieces of asbestos from waste rock. The waste rock is dumped, and the crushed ore is further treated by screens and cone crushers. The large pieces are bagged directly. After this primary treatment, the ore can be stored wet. At some stage, however (previous to the enclosed storage in Figure 2.2), the asbestos must be dried. This stage is the major environmental dust-creating problem in the mining and milling process.

The drying stage uses approximately 80% of the air in an asbestos mill and picks up large amounts of dust. The air temperature is high (60-250°C) and this creates collection problems. Traditionally, cyclones were used to eliminate dust from this process; however, wet collectors, electrostatic precipitators and bag filters are now being used. The dryers themselves are either rotary drums with internal baffles or tower driers.

After the drying stage, the ore is further crushed and screened. Movement is by conveyors and cone crushers and vibrating screens are used. In this stage, the fibers are separated from rock and partly graded on the basis of length. In this stage, also, the fiberizers break up the ore into a more open structure. Repeated milling and screening finally breaks the asbestos down to a stage where the fiber can be air-entrained. It is then air-conveyed to cyclones, where it is collected. The rock left in this process goes to tailings. The air outlet from the process cyclones is cleaned and exhausted or returned. The dust from the cyclone air outlet is screened and bagged as the finest (float) grades of asbestos. The solids from the cyclone bottom proceed to the fiber lines of the process. By this stage they are in several grades.

The fiber lines perform further classification, separation and cleaning. Again, further aspirating enables the opened fiber to be entrained into the the air and collected by cyclones.

Packaging is accomplished by compression and wrapping or by blowing. Shorter fibers are packed more tightly than larger ones. As may be seen from this description, nearly every stage of this sophisticated milling process requires some type of dust control. At all stages, there is a need to enclose the process to minimize the amount of air that must be treated. The asbestos mills in Quebec used 12,000,000 cubic feet per minute (cfm) of air in their processes in 1971.[2]

Exposure

The most complete information on long-term exposure of miners to asbestos dust comes from South Africa. Information on dust levels in these mines is available from 1940 on.[5]

There are two major asbestos-mining areas in South Africa; North West Cape Province (NWC), where crocidolite (Cape blue) is mined, and Transvaal (TVL) where crocidolite (Transvaal blue) and amosite are mined. Chrysotile is also mined in South Africa, but the production is relatively small.

In the 1940s, 1950s and early 1960s, Conimeter counts of particles and fibers of all sizes indicated that both mines and mills in the NWC were, on average, approximately twice as dusty as the TVL mills and mines. The averaged counts ranged from 96 particles per cubic centimeter (ppcc) to 2,687 ppcc. Table 7.1 shows the dust concentrations in the South African mines and mills from 1940-1971.[4]

Table 7.1. Dust Levels in South African Mines[a]

Period	Fiber and Particles per cm^3					
	Crocidolite				Amosite	
	Cape		Transvaal			
	S	U	S	U	S	U
1940-1945	2,687	1,139	—	—	1,458	350
1946-1950	1,742	483	—	—	571	315
1951-1955	1,036	404	—	—	646	445
1965-1966	375	370	633	197	495	130
1968-1969	250	192	246	167	522	161
1970-1971	273	146	110	96	354	150

[a]Counts were derived from konimeter up to 1965, and thereafter by thermal precipitator, and refer to particles and fibers larger than 0.5 μm: S = surface; U = underground; 300 ppcc is equivalent to 10 mppcf.

The authors presenting the above figures warned that they must be assessed with the following points in mind:

1. that dust concentrations causing the observed cases of asbestosis and pleural plaques occurred in the past, but that such heavy concentrations do not exist today;
2. that the dust concentrations referred to were recorded by means of konimeters and thermal precipitators;
3. that preliminary observations in another study tend to indicate that when only *fiber* concentrations are estimated in konimeter and thermal precipitator samples, these must be multiplied by a factor of 1-6 to obtain a correlation with fiber concentrations assessed by collection on filter paper, and the particular factor to be used is determined by the type of fiber involved; and
4. that the safe standard arrived at (~100 total ppcc) represents a long-term mean; thus, the equivalent short-term standard is multiplied by a factor of 1-3, depending on the coefficient of variation of the relevant dust level; hence, if the membrane filter is used, the short-term standard may be as high as 45 f/cc depending on the type of fiber and the coefficient of variation of the dust level.

In Canadian mines, dust surveys have been done since 1948 using the impinger technique.[5] The count, in mppcf, referred to as total dust, including fibers and particles, is counted at a 100-fold magnification and light field illumination. The results are summarized in Figure 7.1.

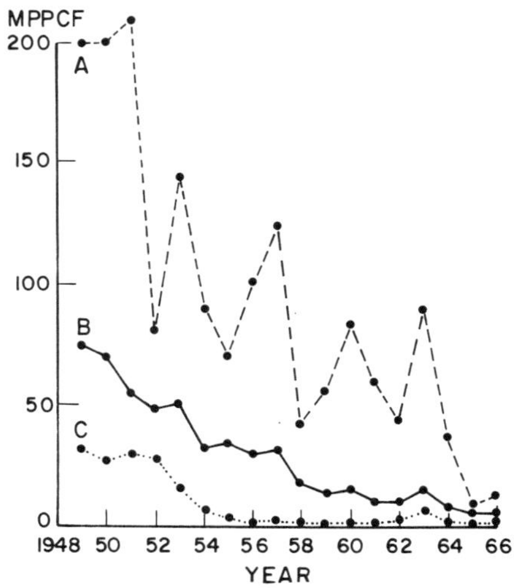

Figure 7.1. Chrysotile dust exposure in Quebec.[6] The mean annual dust concentrations are presented. The mill with the highest measured dust levels in the industry is A; B is the mean concentration for all mills; and C is the mill with the lowest dust levels in mppcf.

Fiber counts based on the membrane filter method have only been applied recently in Canadian mines and mills. As a result, there is little publicly available data concerning fiber levels in these occupational environments. The few data available are limited in scope and given in an indirect fashion. Such indirect evidence[5] would indicate fiber levels ranging from 2 or 3 f/cc to over 50 f/cc, but no idea of time-weighted averages can be deduced.

Long-term exposures in American mines are not documented. However, in 1972, Schutz *et al.*[6] surveyed six asbestos mills, four open pit mines and two underground mines in the U.S., following the recommended NIOSH procedure. Table 7.2 indicates the major results of this study. It may be seen that laborers have the highest recorded exposures in this survey, and the authors suggest that this may have been due to the dry sweeping being done. It should be noted that the highest and lowest exposures tabulated are averages over the mills and mines surveyed. The highest individual count was more than 550 f/cc. The authors concluded that:

1. Some operators are adequately controlling the amount of asbestos fibers in the breathing zone of workmen. However, many are using inadequate control measures and practices which result in workmen being exposed to asbestos fiber concentrations above 5 f/ml.
2. Asbestos or asbestiform fibers being associated with the mining of other minerals complicates the problem of evaluating the health hazards of dusts in mines and mills throughout the United States.
3. It is believed that operators of most mills and mines should be able to meet a "5 fibers/ml of fibers greater than 5 μm in length" standard by following practices such as, but not limited to, those listed below:
   a. installing and using effective and efficient control systems and collectors to prevent dust from becoming airborne;
   b. establishing a rigid maintenance program for process and dust control equipment;
   c. packaging the finished product in airtight containers;
   d. modifying processing machinery or replacing it with equipment designed to achieve adequate dust control;
   e. enclosing all conveyor belts, screens, crushers and other operations, and using exhaust ventilation and collectors to minimize airborne dust;
   f. using vacuum cleaning systems for cleanup work to eliminate airborne dust from sweeping;
   g. following established recommended practices for the use of respiratory protective equipment;
   h. using accepted permissible methods for controlling dust when drilling in pits and mines;

i. implementing methods for keeping dust as stockpiles, roads, millyards and plant areas from becoming airborne; and
j. providing enclosed cabs with filtered air systems for operators of mobile equipment used in mill and mine surface areas.

4. New techniques such as the wet-milling process presently being investigated by some operators should be evaluated as a possible aid in complying with lower standards being considered.

We may assume that these conclusions are applicable to most mining and milling processes.

In general, we can conclude that the drying operation presents the worst problem for controlling environmental dust and that several mill locations and operations present problems for occupational exposures.

Table 7.2. Average of Asbestos Fiber Exposures (daily weighted averages) in Mills and Mines

Occupation	Total Samples	Asbestos f/ml (>5 μm in length) Highest Exposure	Lowest Exposure	Average Exposure
Mill				
Blender, bagger and packer	132	24.8	2.7	11.3
Palletizer and car loader	31	18.1	1.2	10.0
Crusher and dryer operator	32	20.3	7.1	11.0
Foreman	32	15.8	4.3	8.3
Laboratory technician	33	17.4	4.5	8.8
Laborer	37	17.3	8.2	12.1
Maintenance man	50	16.9	3.8	8.4
Mill operator	50	17.9	4.7	11.5
Other	39	10.5	2.3	6.0
Mine				
Driller	31	7.8	1.0	2.8
Explosives man	6	0.6	0.1	0.3
Heavy equipment operator	55	1.7	0.5	1.1
Maintenance man	7	0.5	0.1	0.3
Truck driver	20	0.8	0.2	0.5
Other	21	0.8	0.04	00.5

Control

The mining and milling industries use all aspects of modern dust control technology. Several processes, such as drilling and drying, present unique problems. Wetting as a dust control process has some particular advantages in open pit mining, but product degradation and cold weather render it a less than ideal process. Enormous amounts of air are required for drying and dust control in asbestos mills. The proper treatment of this air before dumping or recirculation requires careful consideration of the economics of control.

ASBESTOS-CEMENT PRODUCTION

Emissions

The wet mechanical process for manufacture of asbestos pipe is shown in Figure 7.2. Asbestos cement products represent the largest single use of asbestos, and pipe is the most important cement product. The process begins at the product tanks (1) where asbestos, silica and cement are mixed. Approximately 15-30% of the final product is chrysotile. Some crocidolite is also used. After electronic weighing (2), the mixture is dry mixed (3). The asbestos adds structural strength to the mixture, and it is important that the fibers be randomly arranged and thoroughly broken up (fiberized or willowed) to accomplish this. The mixture is then wetted and moved through a conveying trough (4) to the wet mix vat (5). The mixture is then taken up by screen cylinder molds (6) and over a vacuum box (7) by a felt roll, during which period water is removed. The felt deposits the slurry onto a mandrel (8) to the desired thickness. The mandrel with the pipe (9) is then moved to where the mandrel can be removed electrolytically (10). The pipe is precured (11) and the mandrel removed. The pipe is labeled (12) and then spends a further period of curing (13). The pipe is next autoclaved (14) under steam. The cured pipes are lathed to remove rough edges (15), tested (16) and stored (17).

Exposure

Since much of this process is wet there is less danger than in many much smaller dry processes. However, there is a high dust exposure during the mixing operation, and a potential dust hazard also arises from dry fragments and shavings on the floor and machinery surfaces. The exposure associated with cutting, sawing and trimming asbestos cement products could be very high if proper dust control measures are not applied.

In seven asbestos cement pipe plants surveyed by the NIOSH, it was shown that the highest mean of 6.3 f/cc was obtained in mixing and the lowest mean of 0.5 f/cc in miscellaneous operation. In three asbestos cement shingle, millboard and gasket plants, the highest count of 4.4 f/cc was obtained in mixing and the lowest of 0.7 f/cc in warehousing.[7] It

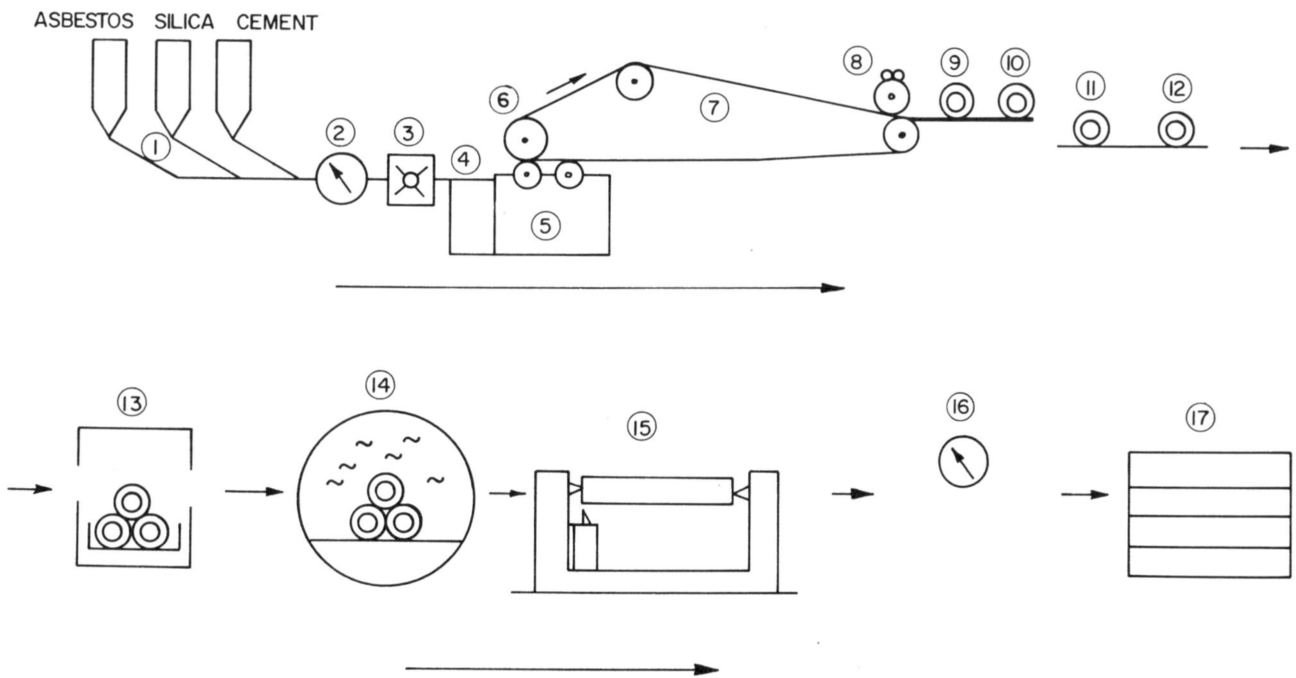

Figure 7.2. Manufacture of Asbestos Cement Pipe:[4] 1. production line bins, 2. electronic scales, 3. dry mixer, 4. conveying trough, 5. wet mix vat, 6. screen cylinder mold, 7. vacuum box, 8. felt deposits stock on mandrel, 9. mandrel (with pipe), 10. electrolytic loosener, 11. slowdown conveyor, 12. stencilling station, 13. air cure room, 14. autoclaves, 15. lathes, 16. testing equipment, 17. storage.

should be noted that this process can expose workers to silica dust as well as asbestos dust.

Control

The process requires control measures during debagging, mixing and machining as well as good cleaning practices. Each of these aspects will be considered in detail later.

INSULATION AND CONSTRUCTION INDUSTRIES

Emissions

Application of asbestos for insulation purposes (especially in spray form) and in removing old lagging materials, commonly in confined spaces, may produce high dust concentrations. The new insulating techniques use little or no asbestos but the exposure problem remains when old lagging is removed. Workmen demolishing buildings may be exposed to the dust of old lagging and other asbestos installation. Refractory bricklayers may be exposed to an asbestos risk as well as a "silica" risk when building, repairing or dismantling furnaces and kilns.

Exposure

The hazards of asbestos spraying are not only limited to the spray operators, but may affect the neighborhood workers, *e.g.,* electricians, plumbers and carpenters. In 1963 the Michigan Department of Health[8] took a series of representative impinger samples at 15-30 ft from the spraying operations and found the dust concentrations ranged from 14-59 mppcf, which was considerably above the then ACGIH TLV of 5 mppcf. The samples of spraying materials when analyzed were found to contain nearly 100% asbestos.

In 1972 Reitz *et al.*[9] indicated that about 40,000 tons of inorganic insulation containing asbestos were used in 1970 by the construction industry as a fireproofing material. They presented Table 7.3 indicating the potential exposures to sprayers.

In five asbestos insulation plants where insulation materials were manufactured, the highest mean of 74.4 f/cc was obtained in mixing and the lowest mean of 16.6 f/cc was in miscellaneous operations. Very high fiber concentrations have been found in naval dockyards, where spraying of insulation and insulation removal were common procedures. Sample data are given by Harries[10] and summarized in Table 7.4.

Table 7.3. Area or Stationary Samples Taken During Spray Operation

Sample	Distance from Nozzle Man (ft)	Count (f/cc)
1	15	17
2	35	10
3	75	46
4	10	71
5	10	70
6	20	37.6
7	20	66.0
Taken 30 minutes after spray operation had ceased for the day.		
8		1.01
9		1.12
10		1.55
11		4.22
Taken 60 minutes after spray operation had ceased for the day.		
12		0.55
13		0.51
14		0.28
15		0.76
16		0.26

Table 7.4. Ranges of Dust Concentrations in Naval Dockyards[a]

Process	
Storerooms	0.1-36
Application of Amosite Sections	9-40
Application and Stitching Asbestos Cloth	0.05-0.26
Removal of Amosite Sections (Boiler Room)	29-1,040
Removal of Blue-Sprayed Asbestos	112-1,906
Removal of Asbestos Acoustic Board	48-683
Bagging Asbestos Debris	106-3,815

[a]Membrane Filter samples taken by Drager Hand Pump, Austen Dymax Pump or Hunt Personal Sampler. Technique of evaluation as described by Asbestosis Research Council, 1964.

Control

In many cases of spraying, ventilation control is not possible. As a result, emphasis has been on use of respirators, hoods and clothing covering. Where possible, of course, asbestos has been replaced in sprayed insulation. It is to be expected by 1978-1979, asbestos compsition spraying on structures will be generally prohibited. This was proposed in 1977 for the United States.

FRICTION MATERIALS

Emissions

Friction materials, such as brake linings and clutch faces, are made by several different processes.[3] Figure 7.3 shows a typical example of the molding of brake linings from a dry mixture. Chrysolite is the main constituent in most friction applications. Friction products also contain large amounts of other materials such as aluminum oxide, brass chips and binders.

The mixture to be used is dry mixed from bins poured into a mold and press-formed under heat. The partially cured result is removed from the mold and further cured. The resulting blank is then cut and ground to the correct shape and the final heat treating process consists of steam-heated forming to the correct shape and baking for a final cure. The resulting output is ground-drilled and counterbored according to the product requirements.

Emissions in a process of this nature can arise from bag opening and dry mixing at the start of the process. Also, associated practices such as weighing, and mixer loading and unloading can cause dust. The larger emissions, however, come from finishing processes such as grinding, sawing, drilling, etc.

Exposure

In five friction plants surveyed by NIOSH, the highest mean of 14.4 f/cc was obtained in a cutting and drilling operation; the next highest of 11.0 f/cc was in mixing, coating and extruding; and the lowest of 2.2 f/cc was in miscellaneous.[7]

Control

Most operations in this field must be controlled by local ventilation since wetting or replacement with other materials is not usually possible.

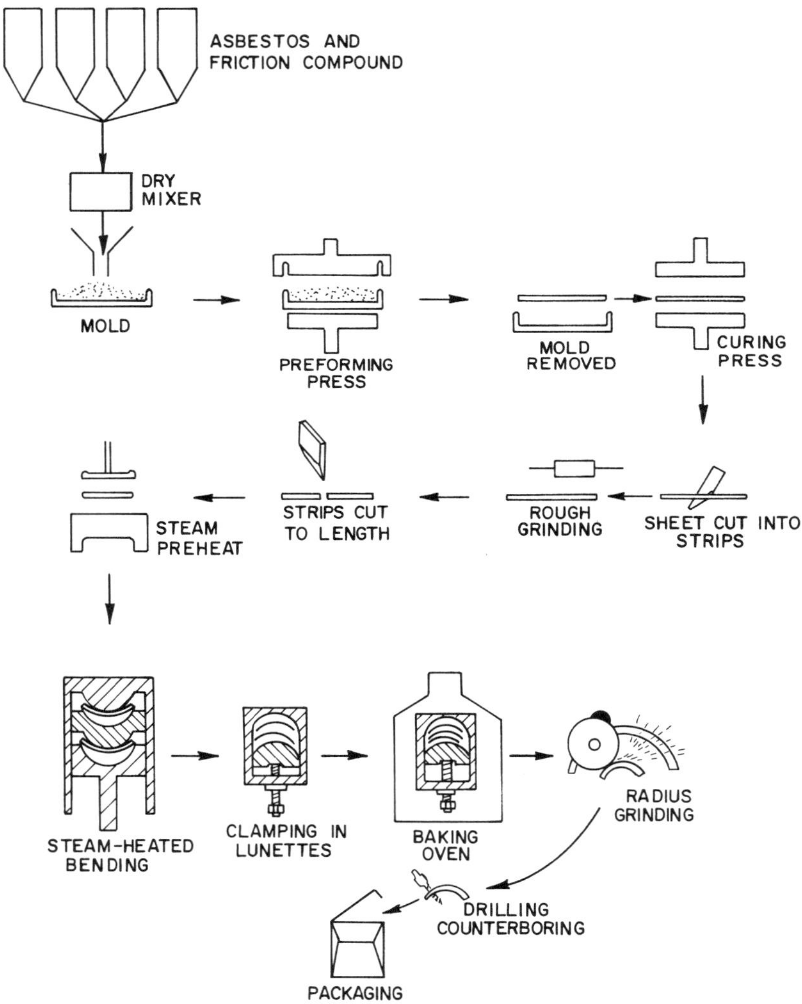

Figure 7.3. Manufacture of dry-mixed molded brake linings.[4]

OCCUPATIONAL EXPOSURE FROM BRAKE SHOES

There has been some concern shown about the asbestos exposure during brake maintenance in the service bays of automotive dealers and garage operators. These garages may service up to 30 cars per day for brakes. Brake lining materials could contain up to 40-60% of asbestos. The sampling data for these types of services are very scanty. Hickish and Knight[11] published some results of the samples taken during the blowing-out procedure, which are summarized in Table 7.5. The authors concluded that although the standard of 2 f/cc may be exceeded during actual brake cleaning, the personal exposures based on the TWA are below the standard.

Table 7.5. Exposure to Asbestos Dust from Brake Shoe Operations

Operation	Type of Samples	No. of Samples	TWA Range (f/cc)
Brake Blowouts	Area	2	1.9 (1.25-2.55)
Car Brake Service	Personal	6	1.9 (0.63-1.12)
Truck Brake Service	Area	6	0.19 (0.07-0.49)
Truck Brake Cleaning and Clutch Repair	Personal	6	2.01 (0.08-7.09)

Much higher levels were found by Rohl *et al.*,[12] where, for example, blowing with compressed air produced fiber counts in the range 7-30 f/cc at 4 ft from the source.

OCCUPATIONAL EXPOSURE FROM FLOOR TILE INSTALLATION

Vinyl-asbestos floor tile contains 15-25% asbestos. Murphy *et al.*[13] studied two installers of floor tile who had no known exposure to asbestos. These workers had frequently sanded vinyl tile floors before installation of new floor coverings. One of the workers had biopsy-proved mesothelioma and the other had extensive pleural calcifications. The authors simulated the work exposures in a room 10 ft x 12 ft x 7 ft exhausted at the rate of six air changes per hour. Two personal membrane samples were worn by an operator sanding the tile for approximately 20 minutes using a conventional belt sander. Dust concentrations for the parallel samples were 1.2 f/cc and 1.3 f/cc of air, respectively. These counts are below the TLV proposed by the NIOSH and others, but do indicate potential exposures. The TLV could have been exceeded if the air changes per hour had been less. The fact that the two installers contracted diseases associated with asbestos is a strong indication that a significant exposure could exist during sanding.

TEXTILE PRODUCTS

Emissions

The various processes involved in manufacturing of asbestos textiles are summarized in Figure 7.4.[14] The dominant form of asbestos used is chrysotile. The asbestos used for textiles has long fine fibers and must be thoroughly fiberized before use. Additives are usually employed to improve spinning characteristics and other fibers, such as cotton, are often added. *Carding* combs the fibers parallel then strips of parallel mat are separated and compressed into strands in the *roving* process. These untwisted strands are spun onto spindles to be twisted to appropriate yarns. Cord may be produced with a series of yarns and either yarns or cord may be woven or braided to make a wide range of products. Much of the machinery used for asbestos textiles is similar or identical to that used in normal textile operations.

As can be seen from the process, most or all of the operations are dry and, as a result, dust may be created at all stages of manufacture. While many operations can work in totally enclosed environments, maintenance and loading operations can still produce the potential for considerable dust.

Exposure

Exposures in a textile mill in Britain have been summarized by Smither and Lewinsohn,[15] and the results are presented in Table 7.6. It can be seen that high levels of fiber are possible in a modern plant.

Control

In addition to dilution procedures, which clean and recycle or exhaust the air in textile plants, extensive enclosure and local ventilation schemes are employed. Often, oils are added to the material to aid in suppression of the dust. Also, manual operations are minimized to the extent possible.

Recently, a new wet process named "Fortex" has been developed[15] for producing chrysotile yarn. The fiberizing has been accomplished by chemical means, and the character of the resulting product is also changed. Very much lower dust levels are encountered in this process (Table 7.7).

OTHER PROCESSES

There are many other asbestos-using manufacturing processes. The ones chosen for detailed discussion were selected both from large operations such as mining and asbestos cement manufacturing and from naturally small processes such as brake shoe and vinyl floor tile work areas. As can be seen, there is

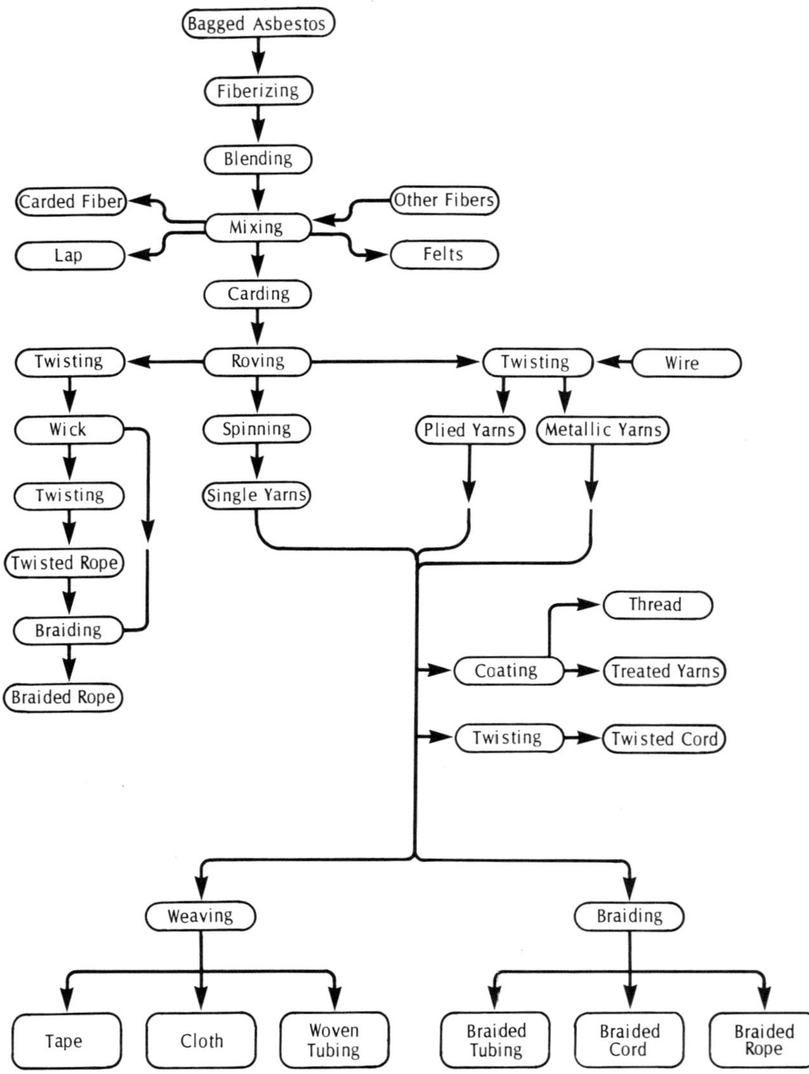

Figure 7.4. Manufacture of asbestos textile products.[14]

Table 7.6. Asbestos Dust Levels in a British Textile Factory, 1961-1971[15]

Department	Process	Yearly Mean Dust Levels (f/ml)			
		Long-Running Thermal Precipitator and Membrane Filter		Static Sampler	Personal Sampler
		1961	1966	1971	1971
Fiberizing	Bag slitting	4.5	4	3	1
	Mechanical bagging	4.0	4.5	4	1
Carding	Fine cards	5.5	5.5	3.5	2
	Medium cards	7.5	8	4.5	3.5
	Coarse cards	7.0	7.5	8	6
	Electrical sliver cards	5.0	2	1.5	1
Spinning	Fine spinning	4	3.5	2.5	3
	Roving frames	5.5	5.5	6	3
	Intermediate frames	5.5	5.5	5.5	3
Weaving	Beaming	8	3.5	0.5	0.5
	Pirn weaving	3	2.9	1.5	1
	Cloth weaving	3	2	2	1
	Listing weaving	2	1	0.5	0.5
Plaiting	Medium plaiting	4	4	4	2

Table 7.7. Dust Levels in Fortex Processing, 1971[15]

	Fine		Medium	
	Static Sampler (f/ml)	Personal Sampler	Static Sampler (f/ml)	Personal Sampler
Forming	0.5	0.5	0.5	0.5
Spinning	0.5	0.5	0.5	1
Winding	0.5	0.5	0.5	1

considerable common ground in these fields if we consider the elemental operations such as mixing, fiberizing, screening, drilling, grinding, etc. As a result, we shall devote much of our effort to illustrating control techniques for these operations. Typical exposures to be expected in normal machining operations are given in Table 7.8.

Table 7.8. Machining Processes Using Cement Sheet and Pipes and Insulation Board[17]

Operation	f/ml
Cement Sheet and Pipes	
Machine Drilling	<2
Hand Sawing	2-4
Machine Sawing Without Effective Local Exhaust Ventilation	
Jig saw	2-10
Circular saw	10-20
Insulation Board	
Drilling Vertical Structures	2-5
Drilling Overhead	4-10
Sanding and Surforming	6-20
Scribing and Breaking	1-5
Hand Sawing	5-12
Machine Sawing Without Effective Local Exhaust Ventilation	
Jig saw	5-20
Circular saw	20 upwards
Unloading Deliveries of Board, Short-Term Sampling	
Cut pieces	5-15
Standard sizes	1-5

Many other manufacturing processes using asbestos are described and discussed in the literature.[3]

REFERENCES

1. Rajhans, G. S. "Fibrous Dust - Its Measurement and Control", *Can. Inst. Min. Metal. Bull.* (August 1970), pp. 900-910.
2. Hutcheson, J.R.M. "Environmental Control in the Asbestos Industry of Quebec," *Can. Inst. Min. Bull.* 64: 83-89 (August 1971).
3. "Control Techniques for Asbestos Air Pollutants," Environmental Protection Agency. AP-117, Reserach Triangle Park NC (1973).
4. Slwis - Cremer, G.K. "Asbestosis in South Africa - Certain Geographical and Environmental Considerations," *Ann. N. Y. Acad. Sci.* 132: 215-234 (1965).
5. Gibbs, G.W., and M. Lachance. "Dust Exposure in the Chrysotile Asbestos Mines and Mills of Quebec," *Arch. Environ. Health* 24: 189-197 (1974).
6. Schutz, L.A., W. Bank and G. Weems. "Airborne Asbestos Fiber Concentrations in Asbestos Mines and Mills in the United States," *United States Bureau of Mines, Technical Progress Report* 72, (July 1973).
7. Health, Education, and Welfare, U.S. Department of Public Health Health Service, Health Services and Mental Health Administration. "Criteria for a Recommended Standard for Occupational Exposure to Asbestos," National Institute for Occupational Safety and Health, HSM #72-10267, Washington, DC (1973).
8. "Construction Health Hazard," *Mich. Occup. Health* 8 3) (1963).
9. Reitz, W.B., W.J. Nicholson, A. Holiday and I.J. Selikoff. "Application of Sprayed Inorganic Fiber Containing Asbestos: Occupational Health Hazards," *Am. Ind. Hyg. J.* 33(3) 178-192 (1972).
10. Harries, P.G. "Asbestos Hazards in Naval Dockyards," *Am. Occup. Hyg.* 11: 135-145 (1968).
11. Hickish, D.E. and K.L. Knight. "Exposure to Asbestos During Brake Maintenance," *Ann. Occup. Hyg.* 13: 17 (1970).
12. Rohl, A.N., A.M. Langer, M.S. Wolf and I. Weisman. "Asbestos Exposure During Brake Lining Maintenance and Repair," *Environ. Res.* 12: 110-128 (1976).
13. Murphy, R.L., *et al.* "Floor Tile Installation as a Source of Asbestos Exposure," *Am. Rev. Resp. Dis.* 104; 577-580 (1971).
14. *Handbook of Asbestos Textiles,* 3rd ed. Asbestos Textile Institute Pompton Lakes NJ (1967).
15. Smither, W.J., and H.C. Lewinsohn. "Asbestosis in Textile Manufacturing," *Biological Effects of Asbestos,* International Agency for Research on Cancer, Lyon (1973).
16. Department of Employment, U.K. "Precautions in the use of Asbestos in the Construction Industry," HMSO London (1974).

CHAPTER 8

NONOCCUPATIONAL EXPOSURE TO ASBESTOS

INTRODUCTION

The risk attached to asbestos dust discharged from asbestos mills and processing factories has been a matter of contention since Wagner in 1965[1] reported cases of mesothelioma in South African children who played on and around asbestos dumps.

Three factors have led to the belief that there are nonoccupational exposures to asbestos:

1. the development of mesothelial tumors in persons who have had no occupational or other exposure to asbestos except residence near a source of asbestos dust;
2. the demonstration of asbestos bodies in the lungs of members of the general public; and
3. radiological evidence, in the vicinity of a mine or large commerical user of asbestos, of the pleural changes commonly associated with asbestos exposure.

ENVIRONMENTAL EXPOSURES DUE TO MINING

The following are the major sources of air pollution as a result of asbestos mining:

1. removal of overburden from ore deposits, shoveling of loosened ore, surface scraping of ore, preliminary screening and conveying of ore at the mine, and loading ore into trucks;
2. loads of ore in transit from mine to processing mill;
3. roadways surfaced with asbestos mill tailings;
4. dumping of ore from trucks at the mill site;
5. effluent from ore dryers;
6. emission from conveyors exposed to the environment;
7. tailing conveyors, placement of tailings onto an existing dump, leveling

of the dump to permit further deposition of wastes, and from direct entrainment of surface dust by ambient air currents; and

8. passive uncovered tailings sites subject to wind erosion.

AIR POLLUTION DUE TO ASBESTOS SPRAYING

Minute amounts of asbestos in urban areas were detected by Selikoff *et al.*[2] when samples were taken near building sites where asbestos spraying was being undertaken. The maximum concentration was 80 ng/m^3. There are, however, sampling problems due to the nature of atmospheric dispersion. The "plume" can be expected to behave in a similar way to a smokestack plume.

ASBESTOS EXPOSURE IN HOUSES, OFFICES AND OTHER BUILDINGS

There is minimal dust exposure from materials incorporating asbestos (*e.g.*, wallboard, ceiling tiles, etc.) in the house, office and other buildings. However, sprayed acoustic and thermal insulation, if not properly sealed, can become friable and constitute a potential longterm hazard to health, especially if the asbestos-covered surface is present in the air supply systems of occupied buildings. It has been the practice frequently (Figure 8.1) to make the region above the false ceiling act as the return air plenum in air conditioning systems. Thus, the air can erode the asbestos sprayed on the exposed surfaces. This exposure could be severe for maintenance workers who must work in confined spaces where asbestos was applied in the past. Figure 8.2 shows an example of this situation where the spray is easily removed, while Figure 8.3 shows the material spalling in a poor installation.

Lumley *et al.*[3] measured asbestos dust concentrations in a group of storehouses insulated with sprayed crocidolite and amosite. Their results showed the range of concentrations varying from 0.26-11.89 f/cc. They concluded that the concentrations of crocidolite and amosite dust that may occur when unsealed sprayed insulation of fallen asbestos debris is disturbed are likely to be in excess of the threshold levels. This is especially so for crocidolite, if the threshold level is to be held below 0.2 f/cc. This study was in contrast to an earlier study by Byrom *et al.*,[4] which indicated no hazard to the health of the occupants of completed buildings. Private communications with others who have made occasional measurements of the office buildings indicate that sprayed asbestos insulation can be a potential hazard to health of the occupants of the buildings if the insulation is unprotected against damage.

Nicholson *et al.*[5] presented a study in 1971 which found that air contamination may exist in public buildings where asbestos materials were used for fireproofing, decorative or acoustic spray applications.

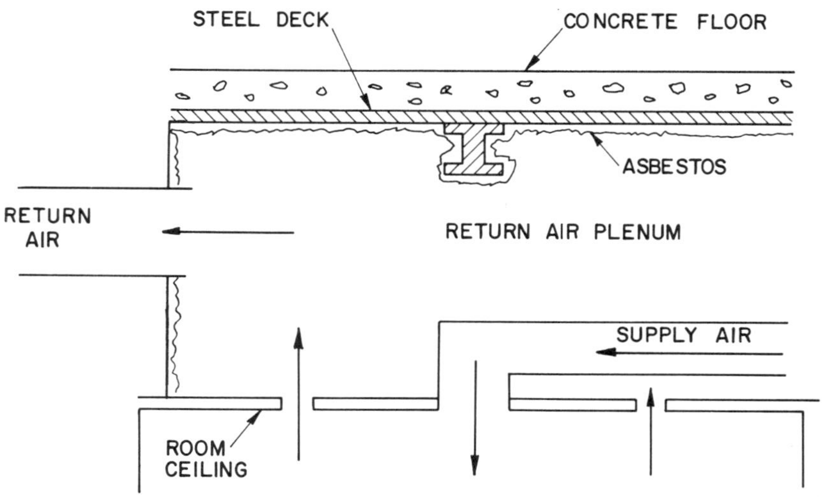

Figure 8.1. Schematic diagram of asbestos-lined return air plenum.

The report states "that half the new (circa 1958-1970) office buildings in the United States were fireproofed with asbestos materials. Millions of people, as a consequence, today could be endangered by the air they breathe while at traditionally nonhazardous jobs." (Because of better knowledge of the health hazard, spray application was largely prohibited by the early 1970s.)

It was also shown[5] that significant fiber counts have only been found where directly visible damage to the sprayed materials was evident. The demolition of asbestos-sprayed highrise buildings could result in the generation of a large amount of asbestos fiber dust.

In 1975 Nicholson and his co-workers produced a further study on the subject.[6] In this work they take care to identify separately the different processes by which asbestos compounds have been applied:

1. Wet Method (cementitious spray)
   In this process, the material (containing approximately 5-30% chrysotile) is mixed as a slurry and a pump sprays the slurry from a nozzle. Portland cement and gypsum provide a cementitious binder, which makes the surface less susceptible to friability.
2. Dry Method (fibrous type)
   In this process, dry material, including asbestos fiber and binders, is dumped into a hopper where the material is dispersed and blown into a hose. The hose has at its nozzle end a ring of water jets converging

Figure 8.2. Sprayed asbestos insulation showing possibility of dust source.

downstream several inches from the end of the nozzle. This wets the dust, and the binders allow adhesion to the surface.

The measurement of the chrysotile was accomplished by taking a collected sample, identifying the chrysotile remaining under an electron microscope by visual means and converting the dimensions of the fibers to a mass-based result. Nicholson *et al.*[6] were then able to show that significant contamination of the air occurred in buildings in which the dry method had been used and that no significant contamination was present in the samples from the wet method. Values of asbestos concentrations ranged from 0-200 ng/m^3. This can be compared with ambient air samples measured by the same procedure of 0-100 ng/m^3 and occupation exposures of 1,000-100,000 ng/m^3. Fiber counts in the study using the NIOSH method were insignificant (< 0.05 f/cc).

Figure 8.3. Spalling of asbestos-sprayed insulation due to improper procedures.

ENVIRONMENTAL POLLUTION FROM BRAKE SHOES

Lynch[7] has shown that there are few or no fibers in the dust produced from bus and automobile brakes. He indicated that the contact temperature between the lining and drum may be higher than the decomposition temperature of asbestos, and the decomposition product of chrysotile asbestos does not include crystalline fibers.

Luxon[8] also indicated that samples from brake drums, when subjected to X-ray diffraction analysis showed between 1% and 3% of chrysotile, although the original lining material contained more than 40% asbestos. He concluded that this is no doubt due to the thermal degradation of chrysotile at the high temperatures generated by friction between the lining and the drum.

However, in New York City, airborne asbestos concentrations in a toll booth were found to be three to five times background levels.[5] It appears that further data are needed before the impact of brake lining decomposition in the environment can be ascertained.

COMMUNITY EXPOSURE FROM ASBESTOS-ASPHALT PAVEMENTS

Asbestos is added to asphalt to obtain an improved pavement overlay with increased cohesion and abrasion resistance. The amounts of asbestos may vary between 2% and 3% by weight. This allows increases in the quantity of asphalt in the mix by 30% to 100% and gives considerably improved properties to pavement overlays. This type of paving has been applied extensively in California, where it is estimated that more than 20% of the population resides in proximity to asbestos-asphalt paving. The procedure is becoming increasingly popular elsewhere, especially on turnpikes.

The asbestos in asphalt is never completely locked in and it is suspected that due to increased vehicular activity some fibers may be separated and are likely to be airborne in due course. In addition, asphalt paving plants can become possible sources of dust if asbestos is being used in the mixing process.

EXPOSURE FROM WASTE DISPOSAL

Asbestos in finished products is combined with other nonfibrous materials in such a manner that once the product is used, asbestos can rarely be separated in usable form. Thus, most scrap containing asbestos becomes part of the general run of solid waste. Incineration releases few fibers, if any, because of thermal degradation of the asbestos.

Generally speaking, asbestos waste falls into the following categories:

1. fine dust;
2. loose fiber, swarf, small offcuts, floor sweepings, etc.;
3. waste material from fixing or stripping of insulating materials containing asbestos;
4. offcuts, broken pieces and reject material from friable materials;
5. offcuts, broken pieces and reject material from high-density products such as asbestos cement, moulded plastic materials, etc.; and
6. empty sacks or bags that have contained loose asbestos fiber.

There is, at present, no indication that there is significant nonoccupational exposure due to waste disposal.

REFERENCES

1. Wagner, J. C. "The Sequelae of Exposure to Asbestos Dust," *Ann. N.Y. Acad. Sci.* 132:691 (1965).
2. Selikoff, I. J., W. J. Nicholson and A. M. Langer. "Asbestos Air Pollution," *Arch. Environ. Health* 25:1 (1972).
3. Lumley, K. P. S., P. G. Harries and F. J. O'Kelley. "Buildings Insulated with Sprayed Asbestos: a Potential Hazard," *Ann. Occup. Hyg.* 14:255 (1971).

4. Byrom, J. C., A. A. Hodgson and S. Holmes. "A Dust Survey Carried Out in Buildings Incorporating Asbestos Based Materials in their Construction," *Ann. Occup. Hyg.* 12:141 (1969).
5. Nicholson, W. J., A. N. Rohl and E. F. Ferrand. "Asbestos Air Pollution in New York City," in *Proc. 2nd Int. Clean Air Cong.,* H. M. Englund and W. T. Beery, Eds. (New York: Academic Press, (1971), pp. 136-139.
6. Nicholson, W. J., A. N. Rohl and I. Weisman. "Asbestos Contamination of the Air in Public Buildings," EPA-450/3-76-004, National Technical Information Service #PB-250 980 (October 1975).
7. Lynch, J. R. "Brake Lining Decomposition Products," *J. Air Poll. Control Assoc.* 18:824 (1968).
8. Luxon, J. "Technical Implementation of the New Asbestos Regulations," *Ann. Occup. Hyg.* 13:23-24 (1970).

CHAPTER 9

ASBESTOS IN WATER AND CONSUMER PRODUCTS

INTRODUCTION

The recent discovery of submicron asbestos fibers in beer, wine and drinking water has resulted in public concern that a health hazard exists. It is also suspected that the gastrointestinal (GI) tract receives most of the inhaled and subsequently ingested fibers that may cause GI cancer.

IN CITIES

River and lake water near asbestos ore bodies has always contained asbestos fibers of microscopic dimensions. Increased industrial applications of asbestos have increased the number of rivers and lakes with asbestos and, of course, the quantity. One of the most well-known examples is in Duluth, Minnesota. About 20 years ago the Reserve Mining Company (jointly owned by Armco and Republic Steel) began processing taconite, a low-grade iron ore, on the lakeshore, 60 miles northeast of Duluth, dumping the mining wastes, or tailings, into the lake. To date, the company has discharged more than 200 million tons of tailings.[1]

Local ecological groups began protesting some years ago. A suit against the mining company was eventually brought by the U.S. Government, joined by the states of Minnesota, Wisconsin and Michigan, the Environmental Defense Fund and four local ecology groups. The case came to trial in August 1973, some 17 years after the dumping began. The suit charged that the Reserve Mining tailings were not only altering the ecology of the lake but also were being carried southwest by lake currents, so that they ended up in the drinking water consumed by the residents of the Duluth area.

The company had long insisted that its tailings were not toxic. Actually, the tailings are composed in large part of cummingtonite and similar minerals.

Cummingtonite is an asbestiform mineral of the amphibole group. Amosite, which is not a true mineral but rather an acronym (*A*sbestos *M*ines *o*f *S*outh Africa), contains cummingtonite.[2] Subsequent tests by EPA chemist Phillip Cook showed that Duluth's drinking water does, in fact, contain asbestos-like fibers in significant quantities. Independent studies by the EPA and the state of Minnesota later confirmed that a substantial quantity of those fibers were asbestos.

According to Cunningham and Ponterfract,[3] asbestos in some Ontario cities ranges from a low of 0.5×10^6 f/l to 3.87×10^6 f/l and approximately 0.0001-0.05 parts of asbestos per billion parts of water. The tap water for Thetford Mines in Quebec has shown 172×10^6 f/l.

IN BEVERAGES

Asbestos fiber has also been found in beer, soft drinks and fortified wines, mainly because pads containing asbestos are used for filtration in the beverage industry.[4] Some values of asbestos fiber in various liquids are given in Table 9.1.

Commercial production and applications of asbestos fibers are usually considered to be the only significant sources of asbestos fibers entering the environment. However, on closer examination it becomes evident that we should consider not only the question of impurities in asbestos, but also that asbestiform minerals are an ubiquitous impurity in many deposits of such commercially valuable nonmetallic minerals as mica and talc. Talc, particularly, is a mineral product with widespread commercial and cosmetic applications.

IN THE ENVIRONMENT

Windom *et al.*[5] investigated the distribution of talc in the atmosphere and in glacier and snow samples to study the migration of pesticides. Their samples covered a worldwide geographic distribution. In practically every sample, amphiboles were detected along with the talc, as might be anticipated from the common occurrence of amphiboles in talc. Cralley[6] has investigated 22 cosmetic talcum products. All had significant fiber contents ranging from 8-30% by count of the total talc particulates, and averaging 19%. The fibrous talc included tremolite, enthophyllite and chrysotile. They made special note of the fact that cosmetic talcum products should be included as a source of fibers from which may be derived ferruginous bodies observed in the lungs of humans.

In another investigation by the U.S. Public Health Service of the source and identification of respirable fibers, Cralley[6] noted that there are more than 100 different natural minerals with some degree of fibrous structure that may occur in respirable sizes. In addition to the asbestos minerals, these included fuller's earth, zeolite, vermiculite, calcium carbonate, gypsum, pyrophyllite, talc, kyanite, hornblends, mica, magnesite and many others.

Table 9.1. Asbestos Fibers in Beverages and Water[3]

Sample	Source	Millions of Fibers per Liter
Beer	Canadian 1	4.3
Beer	Canadian 2	6.6
Beer	USA 1	2.0
Beer	USA 2	1.1
Sherry	Canadian	4.1
Sherry	Spanish	2.0
Sherry	South African	2.6
Port	Canadian	2.1
Vermouth	French	1.8
Vermouth	Italian	11.7
Soft Drink	Ginger ale	12.2
Soft Drink	Tonic water I	1.7
Soft Drink	Tonic water II	1.7
Soft Drink	Orange	2.5
Tap Water	Ottowa, Ottawa River (F)[a]	2.0
Tap Water	Toronto, Lake Ontario (F)	4.4
Tap Water	Montreal, St. Lawrence River (F)	2.4
Tap Water	Hull, Quebec, Ottawa River (NF)[b]	9.5
Tap Water	Beauport, Quebec, St. Lawrence River (6 km below Quebec City) (NF)	8.1
Tap Water	Drummondville, Eastern Townships, Quebec, St. François River (F)	2.9
Tap Water	Asbestos, Eastern Townships, Quebec, Nicolet River (F)	5.9
Tap Water	Thetford Mines, Eastern Townships, Quebec, Lac à la Truite (NF)	172.7
Melted Snow	Ottawa, top 30-cm (2-3 weeks precipitation)	33.5
River Water	Ottawa River, at Ottawa	9.5

[a]F: filtration plant used.
[b]NF: no filtration plant used.

REFERENCES

1. "Asbestos in Water," *Lancet,* 2:1256 (1973).
2. "Selected Silicate Minerals and Their Asbestiform Varieties," U.S. Bureau of Mines Information Circular IC8751 (1977).
3. Cunningham, H. M., and R. Ponterfract. "Asbestos Fibers in Beverages and Drinking Water," *Nature* 232:332-333 (July 1971).
4. "Examination of Fibers in Beer," *Nature* 219:93-94 (July 6, 1968).
5. Windom, H., J. Griffin and E. D. Goldberg. "Talc in Atmospheric Dust," *Environ. Sci. Technol.* 1:923-926 (1967).
6. Cralley, L. J. "Identification and Control of Asbestos Exposures," *Am. Ind. Hyg. Assoc. J.* 32(2) (February 1971).

PART II

ASBESTOS DUST: CONTROL

In this section we discuss ways to eliminate and control the dust in various environments, ways to protect personnel when dust cannot be controlled and the economics of control.

CHAPTER 10

NONVENTILATION CONTROL PROCESSES

INTRODUCTION

Asbestos dust arising from the multitude of processes outlined in Part I must obviously be controlled to the greatest extent possible. Local ventilation and control of the air around dust-creating operations is unquestionably the most widely used of all control operations. It is, however, an expensive process and not always totally effective. As a result, we shall begin our discussion of control by looking at the nonventilation processes. These are frequently cheaper and more effective in places where they can be used.

HISTORY OF CONTROL

The first mention of dust control measures was made in the Asbestos Regulation of 1931 in the U.K.[1] However, the regulation only applied to making fiber from ore, mattresses, textiles and insulating slabs, and to the mixing of dry asbestos for more than eight hours per week.

Leathart[2] criticized the regulation. He indicated that the insulating industry was particularly important because many certified cases of asbestosis were derived from it. But although attention was drawn to the insulating industry, it is clear that wherever asbestos or asbestos products are handled, sewn, filed, polished or machined, there should be rigid preventive precautions. These should take the form of total enclosure, dust suppression or extraction, personal respirators, etc. An ideal but unrealistic solution would be the abandonment of all further use of asbestos. In May 1970 the Asbestos Regulation 1969[3] came into force promulgating rigorous standards for preventive measures.

In the United States, Part 1910.1000 of the *Code of Federal Regulations* now defines conditions of exposure for various substances, and Part 1910.1001 specifies standards for asbestos. In various Canadian provinces, standards for application of enclosures and ventilation methods, wearing of protective clothing and respirators have been laid down by the regulation. Most other western countries have developed or are developing standards similar to American or British procedures.

SUBSTITUTION

There is no doubt that the problem of asbestos exposure is solved if a safe, acceptable substitute for asbestos can be found and used. However, it is not always possible to find an effective substitute. In some cases, the type of asbestos should be carefully selected. For example, the less harmful chrysotile might be used to replace crocidolite, but the properties of various forms of asbestos are not the same. Chrysotile has good heat resistance, but poor acid resistance, while crocidolite has good acid resistance, but tends to fuse at high temperature. Thus, while there have been considerable difficulties in finding substitutes, extensive efforts have produced a number of substitutes (Table 10.1).[4]

Fibrous glass, rock wool, mineral wool and ceramic fibers are still the most commonly used substitutes for asbestos. At present, these appear to be less hazardous than asbestos. However, the hazards associated with these substitutes may appear negligible only because they have not been investigated in detail. It should also be realized that the use of substitutes should be controlled in a way that the airborne concentrations do not exceed the TLV of 10 mg/m^3 for nuisance dust. This is especially true where loose fillers containing no asbestos are used (*e.g.,* vermiculite, expanded perlite and diatomaceous earths).

In those plants where immediate substitutes cannot be found (*e.g.,* the product is subjected to long periods of intense heat), an attempt should be made to use locked-in asbestos products, which by nature reduce and in some cases virtually eliminate, the asbestos dust hazards. Examples of these treated asbestos products are graphite-bearing lubricated foil wrappers and gland packings, rubberized sheet, and plasticized or aluminized cloths. The possibility of asbestos substitutes in the asbestos cement and friction products field has been discussed by Green and Pye.[5]

In asbestos cement products, the asbestos acts as a reinforcing fiber. It has high strength and a large aspect ratio. Both these properties produce desirable properties in the final composite. Crocidolite is used in piping because of its greater strength (Table 3.1, p. 26) and its wetting properties. The combined properties of asbestos, especially its wettability, make it easy to form into a slurry during the manufacturing process and to form complex shapes by molding.

In locations where the unique properties of asbestos are not required, quite ordinary replacements are often effective. For cladding purposes, glass-reinforced plastic (GRP) can be used although GRP does not have high fire resistance

Table 10.1. Asbestos Substitutes

Asbestos-Containing Materials	Asbestos-Free Alternatives	Comments
Moulded Asbestos (Amosite Asbestos Bonded with Sodium Silicate)	None	
85% Magnesia (About 15% Asbestos)	Asbestos-free magnesia	Ceiling temperature 316° C; one manufacturer only; colored pink to distinguish from asbestos-containing materials.
Calcium Silicate (About 15% Asbestos)	Asbestos-free calcium silicate	Ceiling temperature varies, generally 600° C.
High Temperature Insulating Block (About 15% Asbestos with Diatomite)	Vermiculite alumina cement block	Some calcium silicates may be suitable alternatives.
Plastics Compositions	Asbestos-free versions of magnesia and calcium silicate	Generally satisfactory for gap-filling and small areas.
Sprayed Asbestos and Asbestos Mineral Wool	1. Sprayed mineral wool 2. Sprayed ceramic fibers 3. Sprayed vermiculite cement and sprayed concrete	1. At development stage only 2. Limited experience, expensive 3. For fireproofing
Self-Setting and Hard-Setting Finishing Cement (up to 40% Asbestos)	1. Asbestos-free finishing cements 2. Plastics finishes applied in paste form and reinforced with glass cloth 3. Sheet finishes (metal, plastics, roofing felt, etc.)	1. May be subject to more cracking than asbestos-containing finishes 2. Colored if required. Less robust and generally more expensive than cement 3. Cheap for straight pipe lengths but expensive where bends and shapes involved
Board and Sheet	Ceramic fiberboard and sheet	Higher ceiling temperature, 1,260° C, than asbestos material; expensive
Rope	1. Glass fiber rope 2. Ceramic fiber rope 3. Silica fiber rope 4. Ceramic fiber felt strip	1. Ceiling temperature, 500°C 2. Ceiling temperature, 1,260° C; expensive 3. Ceiling temperature, 982°C; very expensive 4. May be cheaper alternative to ceramic rope in some cases

Table 10.1, continued

Asbestos-Containing Materials	Asbestos-Free Alternatives	Comments
Cloth, Tape, String, yarn, etc.	1. Glass fiber fabrics and yarn 2. Ceramic fiber and yarn 3. Silica fiber and yarn	1. Ceiling temperatures up to 500°C depending on type of glass. Some cloths may be unpleasant to handle. 2. Ceiling temperature, 1,260°C; expensive 3. Ceiling temperature, 982°C; very expensive

NOTE: Manufacturers have incorporated organic fibers in some of the materials that previously contained asbestos. Such materials may no longer be suitable for certain duties, *e.g.*, where contamination with powerful oxidizing agents is possible.

and combustion can produce toxic fumes. Cast iron sewage pipe is, of course, an alternative to asbestos pipe.

One attractive possibility, which is at present much more expensive than cement pipe, is glass-reinforced cement. Normal glass fiber used in GRP cannot be used since the alkalinity of the cement degrades the fiber. However, alkali-resistant glasses have been developed and are available. The mechanical and fire resistance properties are similar to asbestos pipe and, in some cases, better. The manufacturing process for this product is presently less economical than asbestos pipe and, as a result, the product is considerably more expensive. Glass-asbestos-cement mixtures have also been developed. At present, the cost prevents glass-reinforced cement from achieving wide application.

In friction applications, the detailed functioning of asbestos in the composite is not clear. However, the fibers strengthen the brake pad or clutch face both at low and high temperatures. Also, the friction and wear characteristics of chrysotile are assumed to be responsible for the good performance of the matrix.

Attempted alternatives include sintered metals, ceramics and metallized bonded composites. Use of metals and other high thermal conductivity products, however, creates the possibility of overheating the brake fluid. Carbon fiber-carbon composites have been attempted with some success but are expensive. A rather more successful approach has been to use steel and glass fiber composites. Both tend to be noisy and, on occasion, undependable. Extensive research is underway in attempting replacements in this field.

The spraying of fireproofing and insulation composites containing asbestos is presently either dying out or being forbidden by legislation in most western countries. The mixtures described in the section entitled *Asbestos Exposure in Houses, Offices and Other Buildings,* p. 88, are being replaced with spray mixtures containing vermiculite, mineral wool and ceramic fibers. Many of these mixtures are proprietary and their exact composition is not known; howevery, all are advertised as being asbestos-free and all are fire-retardant in varying degrees.

Among the most valuable characteristics of asbestos compounds are their heat-resistance and fire-retarding properties. In considering alternatives to asbestos, Figure 10.1 may be considered as a partial guide to the thermal operating range of some other materials.

WETTING

Wet processing, as noted before, considerably reduces the possibility of dust creation. Those operations that are amenable to wetting (*e.g.,* asbestos cement production) typically have much lower dust levels than processes that must be run entirely dry (*e.g.,* asbestos textile manufacturing).

Wetting has been used in mines for many years. Dust control is achieved by drilling wet and wetting down broken ore and rock before moving the material to ore raises. The moist conditions of the mine workings tend to allay any dust released in blasting. However, because of extremely cold climates, the Canadian asbestos mines have avoided the use of wet drilling to control the dust and have opted for dry drilling techniques with individual bag collectors installed on each drill. During dry seasons, road dust can be suppressed in open pit operations by water trucks spraying the road surface.

Under certain circumstances, the effect of wetting can be greatly enhanced by a wetting agent.[6] There are a large number of wetting agents available that are commonly used in proportions of around one part per thousand in water. Adding a wetting agent alters the surface tension of water, and, as a result, dust can penetrate into a droplet rather than just adhere to its surface, fine particles are more easily cemented into large agglomerates and sprays form smaller droplets with larger surfaces. As a result, dust capture capability can often be increased many times. While not suitable in applications in which the product would be altered by the presence of the wetting agent, there are several applications to be considered. In mines, the wetting of roads can be done with wetting agents. It is also possible to add agents to water used when removing insulation and performing certain types of cleanup and maintenance operations. Under certain circumstances, surface active agents can be used on the surfaces of exposed stockpiles.

Wetting processes have been developed that greatly reduce the dust arising from the stripping of thermal insulation.[4] Since this is a type of exposure known to give very high and very harmful exposures in the past, the method is extremely useful. Special rubber hose and spray nozzles have been devised that can be

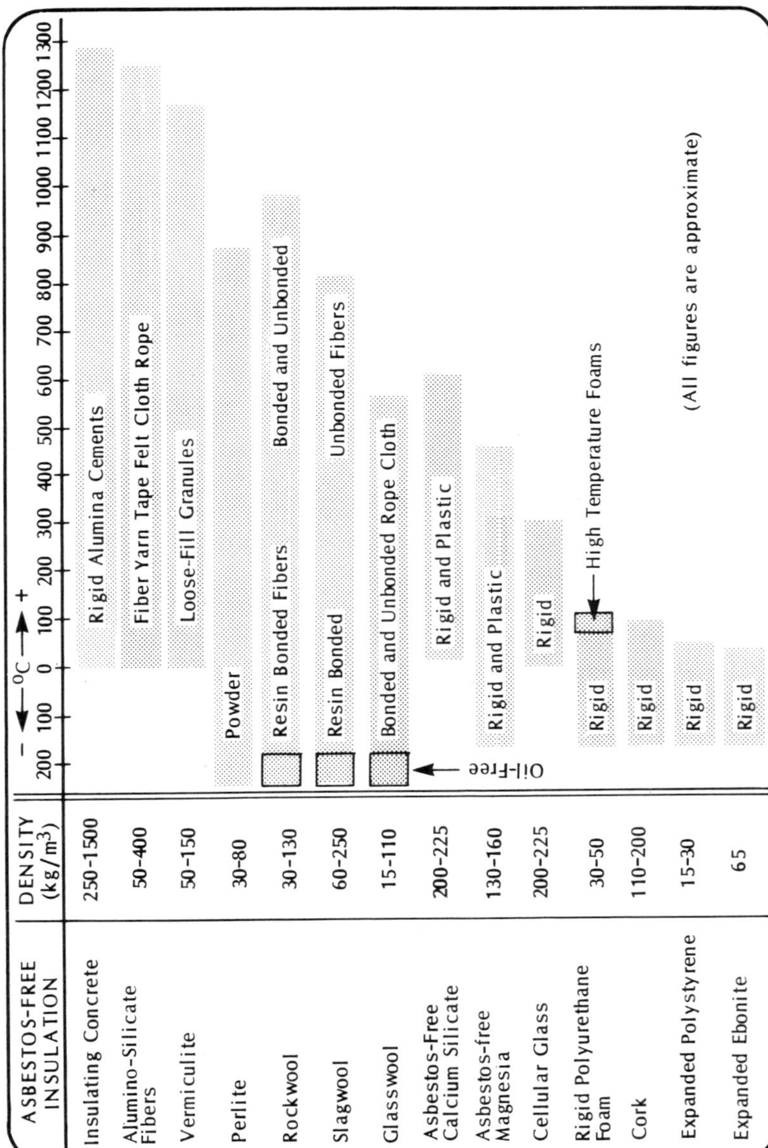

Figure 10.1. Thermal properties of some asbestos substitutes.[4]

inserted in the lagging in such a manner that the asbestos layer is thoroughly soaked before it is removed. A suitable wetting agent may be added to the water to assist penetration. Table 10.2 below shows the effectiveness of wet methods.[7]

Table 10.2. Wet Method Effectiveness of Controlling Airborne Asbestos Fibers

Task	No. of Samples	Fibers (5 μm/cm^3)
Removing Dry Scrap Insulation	13	0.3-13.8
Removing Wet Scrap Insulation	6	0.0-0.3

Wet machining processes are possible in many cases. The procedure requires, of course, some attention to electrical safety and other operational problems associated with water in the presence of lathes, drills, saws etc. The authors are aware of a small plant where large amounts of free running water are run over machining processes such as drilling and lathe cutting. The process seems, in this case, to have produced quite significant reductions in fiber levels. In addition, the floor of this small plant is sluiced at approximately half-hour intervals with water, and most housecleaning processes are wet. Wet sawing also can be used to remove insulation if a reciprocating saw is used. Some asbestos materials for insulation can be wetted effectively within their bag containers before they are applied to the pipes.

In certain textile processes, moistening the asbestos which is being processed reduces the dust production considerably. It has been shown[8] that a wet method sometimes must be augmented by a certain measure of exhaust ventilation or by the provision of dust respirators.

PACKAGING

Serious problems in warehouses, transportation, and loading and unloading operations may be avoided if proper packaging is used. The old method of pneumatic bagging was itself a dusty operation and the hessian bags used allowed dust to escape during transportation and warehousing. Modern bags are effectively airtight if not broken. These bags should be handled carefully so that opening only occurs at proper unbagging stations.

REFERENCES

1. "Industrial Health Legislation: Court Decisions: Workmen's Compensation and Insurance," The Asbestos Industry Regulations, Statutory Rules and Orders, 1931, 14(7):191 (1932).
2. Leathart, G. L. "Asbestos–A Medical History of the 20th Century," *Occup. Health* (1965), pp. 119-131.
3. Department of Employment and Productivity. "Asbestos: Health Precautions in Industry," *Health and Safety at Work 44,* HMSO, London (1970).
4. "Recommendations For Handling Asbestos," *Engineering Equipment Users Association Handbook No. 33,* EEUA, London (1969), pp. 40-41.
5. Green, A. K., and A. M. Pye. "The Search for Asbestos Substitutes," *Asbestos* (May 1977), p. 6.
6. Johnson-March Corp. "Air Pollution Control Through Wet Dust Suppresion," Paper #72-B-23, Annual Meeting of American Institute of Mining Engineers, February, 1972.
7. Fontaine, J. H., and D. M. Trayer. "Asbestos Control in Steam–Electric Generating Plants," *Am. Ind. Hyg. Assoc. J.* 36(2):126-130 (1975).
8. Rajhans, G. S. "Fibrous Dust–Its Measurement and Control," *Can. Inst. Min. Metall. Bull.* 63:900-910 (1970).

CHAPTER 11

LOCAL VENTILATION

INTRODUCTION

The basic control method for asbestos dust is by ventilation. The process assumes that dust will be dispersed into the air by various mechanisms. As a result, it attempts to remove the contaminated air and to clean it without exposing personnel to the contaminant.

BASIC PRINCIPLES

It is obvious that the amount of air contaminated by a dust-creating process must be kept to a minimum. As a result, *local ventilation* is at all times preferable to *dilution ventilation.* The term local ventilation refers to those devices that collect the dusty air as soon as possible after the operation creating the dust. In this way, the amount of air to be treated by the dust system is minimized and the later cleaning process is both cheaper and more easily accomplished. Dilution ventilation, on the other hand, assumes that the entire work area is contaminated and replaces the air in that work area with clean air. Dilution ventilation is usually only economical if the exhaust air need not be cleaned. In asbestos applications, both local and dilution ventilation are frequently employed and, where possible, total enclosure under negative pressure is considered.

The nature of a process and the mode of dust generation determine the general design of an enclosure and local mechanical exhaust provided for. The other guiding factors are: (1) the positioning of the operator in relation to the dust source; (2) air velocity into the enclosure, and (3) the speed at which the dust is released.

If the processes cannot be enclosed, the hoods should be provided as near as possible to the source of the dust. These hoods are to be designed to receive and remove the dust-laden air before it pollutes the general atmosphere. Dust collecting systems associated with hoods must be capable of removing the air exhausted by the hood and the fan must be capable of providing a required *capture velocity* for asbestos dust particles. Thus, the basic principles of local ventilation fall into the folowing categories: selection and determination of air flow distribution and enclosures, local exhaust, air conveying, air cleaning, fan selection, and testing and monitoring techniques.

These elements can be seen in Figure 11.1. The process is hooded, and an inward air flow at the open side of the hood (face velocity) is maintained. The air flows through a ducting system to a precleaning cyclone, which removes the large dust particles only but serves to decrease the loading on the bag filter. The bag filter cleans the dust from the air and the air flows through the fan, which drives the entire system. The cleaned air is then ducted to the outside past a dust monitoring device. Some of the elements shown in this diagram are optional depending on the particular situation; however many other elements may be added.

To discuss in detail the design of these systems, we shall now define some basic air flow quantities. The examples are given in the foot-pound system but a conversion table for the most common quantities is given in Appendix II. The material to be used is described more extensively in the literature,[1-3] and we shall assume here either a familiarity with the more basic material or access to these more general references.

The conservation of mass equation for an incompressible fluid becomes a conservation of volume equation which can be written:

$$Q = VA \tag{1}$$

where Q is the volume flow rate in cubic feet per minute (cfm) or cubic feet per second (cfs)
V is the average air velocity in fpm or fps
A is the cross-sectional area *normal to the flow* in ft^2

While the flow is assumed incompressible in the usual case, this does not imply that density need be considered constant from one use to another for Equation 1 to be used.

The density of air, d, varies with temperature and pressure, and the density of dusty air may be calculated taking the density of the dust into account. However, in ordinary dust loadings such as those found in asbestos ventilation systems, the error is less than 2% (for loadings of 10 gr/ft^3 or less) if the dust is ignored. Much larger density variations occur due to temperature variations. The standard value of density assumed for room temperature air is 0.075 lb/ft^3 (accurate for dry air at 70° F and 29.92 in Hg). Density expressed as lb/ft^3 is, formally speaking, a specific weight. However, the universal terminology in air handling is to evalute density as weight per unit volume.

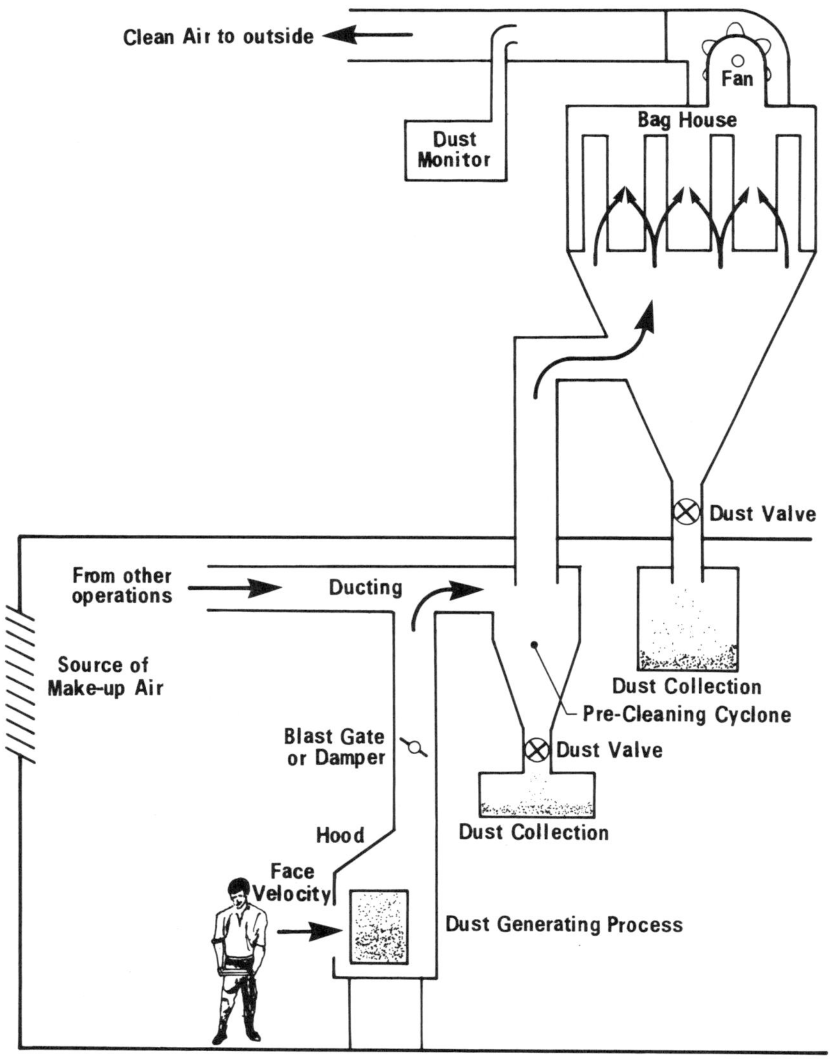

Figure 11.1. Schematic diagram of a basic local ventilation system.

The conservation of energy equation applied for an incompressible fluid with no losses and disregarding elevation changes gives the relationship along a streamline, duct or flow passage between static pressure and velocity. At a point in the flow, the relationship between total pressure, velocity head and static pressure is conventionally written as:

$$TP = SP + VP \qquad (2)$$

where TP is the total pressure at a point in in. of water
SP is the static pressure at the same point in in. of water
VP is the velocity pressure at this point in in. of water

An alternate form of this relationship can be written in terms of feet of fluid flowing (in this case air) as:

$$H = \frac{P}{d} + \frac{V^2}{2g} \qquad (3)$$

where H is the total head in ft of air

P is the pressure in lb/ft^2
d is the density in lb/ft^3
V is the velocity in fps
g is the gravitational constant ($=32.2\ fps^2$)

Each term in Equation 3 is larger than the equivalent term in Equation 2 by the ratio of densities of water and standard air and the factor 1/12 since the terms of Equation 2 are always expressed in inches. Hence, to relate the velocity term in both equations:

$$VP \times \frac{\text{density water}}{\text{density air} \times 12} = \frac{VP \times 62.4}{0.075 \times 12} = \frac{V^2}{2g} = \frac{V^2}{64.4}$$

or

$$V^2\ (\text{fps}) = 4465\ VP$$

and

$$V\ (\text{fpm}) = 60 \times \sqrt{4465\ VP}$$

Hence

$$VP = \left[\frac{V}{4005}\right]^2 \qquad (4)$$

where V is expressed in fpm and VP in in. of water.

If the temperature of the air varies by more than about 50°F, Equation 4 should be altered by a temperature correction as:

$$VP = \left[\frac{V}{4005}\right]^2 \left[\frac{530}{460+T}\right]^2 \qquad (4a)$$

where T is the temperature in °F.
If the gas is other than air, a density correction is required as

$$VP = \left[\frac{V}{4005}\right]^2 \left[\frac{\text{density of gas at design condition}}{\text{density of standard air}}\right] \qquad (4b)$$

The various forms of equation 4 give the conversion factor between air velocities in feet per minute and the equivalent head in inches of water. The other terms in Equation 2 are commonly measured directly in inches of water in ventilation practice.

In an ideal or frictionless fluid, the total pressure is constant along a duct so that for no losses:

$$SP + VP = TP = \text{constant} = TP_0 \tag{5}$$

at all points along the duct. In real flows with friction, however, there are losses of energy or total pressure that must be taken into account. Hence

$$TP_1 = TP_2 + \text{losses}$$

or

$$SP_1 + VP_1 = SP_2 + VP_2 + \text{losses} \tag{6}$$

where point 2 is downstream of point 1. These losses are usually expressed as proportional to the velocity pressure. This does not imply that the velocity head in a system is lost due to friction. In fact, in many geometries, static pressure is lost and velocity pressure is constant. There are two classes of losses–those due to friction in the duct runs and those due to various elements in the flow such as hoods, bends, dampers and filters. The latter are commonly called *minor losses* in fluid mechanics terminology even though their sum may be many times greater than the friction losses in a particular case.

For the friction losses, the magnitude can be evaluated by the Darcy-Weisbach equation:

$$h'_f = f \frac{L}{D} \frac{V^2}{2g} \tag{7}$$

where h_f is the head loss due to friction in ft of air
f is a friction factor dependent upon the Reynolds number, Re, and the surface roughness of the material used for duct construction
L is the duct length in ft, and
D is the duct diameter in ft for circular ducts.

The Reynolds number is given as:

$$Re = \frac{VD}{\nu}$$

where ν is the kinematic viscosity of the gas in f^2/s. The kinematic viscosity of standard air is 1.6×10^{-4} f^2/s. ν is highly dependent on the gas temperature. Equation 7 is usually converted into units consistent with inches of water and this becomes:

$$h_f = 7.5 \times 10^{-7} f \frac{L}{D} V^2 \qquad (8)$$

where h_f is now expressed in inches of water
f is the friction factor
L is length of the duct in ft
D is the duct diameter in in.
V is now expressed in fpm.

It is common practice to assume a value for f of 0.27 for ventilation calculations. This value assumes galvanized ducting with approximately 40 joints per 100 ft of run and a small amount of dust on the wall. The importance of Equation 8 is such that it is frequently given in graphic form. An example[4] is given in Figure 11.2.

Equation 8 has been written for circular ducts where D is the duct diameter in inches. For other shapes of duct, such as square or rectangular cross section, the loss is also calculated from Equation 8, but a hydraulic or equivalent diameter is used. The most common form is:

$$D' = \frac{4A}{P} \qquad (9)$$

where A is the cross sectional area of the duct
P is the perimeter of the duct

D' may be used in place of D in Equation 8 or in the formula for Reynolds number. The error in using this approximation is generally less than 10%

Minor Losses

As previously stated, these losses are ordinarily given as a fraction of VP. Hence, a loss coefficient, K, can be defined by:

$$h_e = K \left[\frac{V}{4005}\right]^2 = K.VP \qquad (10)$$

where h_e is the loss in inches of water due to the loss causing element
K is the loss coefficient which must ordinarily be tabulated based on experimental values

It is frequently difficult to find values of K for typical geometries of ventilation hoods and covers. Extensive sets of values of these quantities are tabulated in the *Industrial Ventilation* manual.[2] The ASHRAE Manual[1] lists losses for many standard air conditioning geometries, and a wide range of unusual configurations are tabulated by Miller.[5]

In summary, then, the balance of pressures along a duct is related by Equation 6, which may be written as:

$$SP_1 + VP_1 = SP_2 + VP_2 + \Sigma(7.5 \times 10^{-7} f \frac{L}{D} V^2)_i + \Sigma \left[K\left(\frac{V}{4005}\right)^2\right]_j \qquad (11)$$

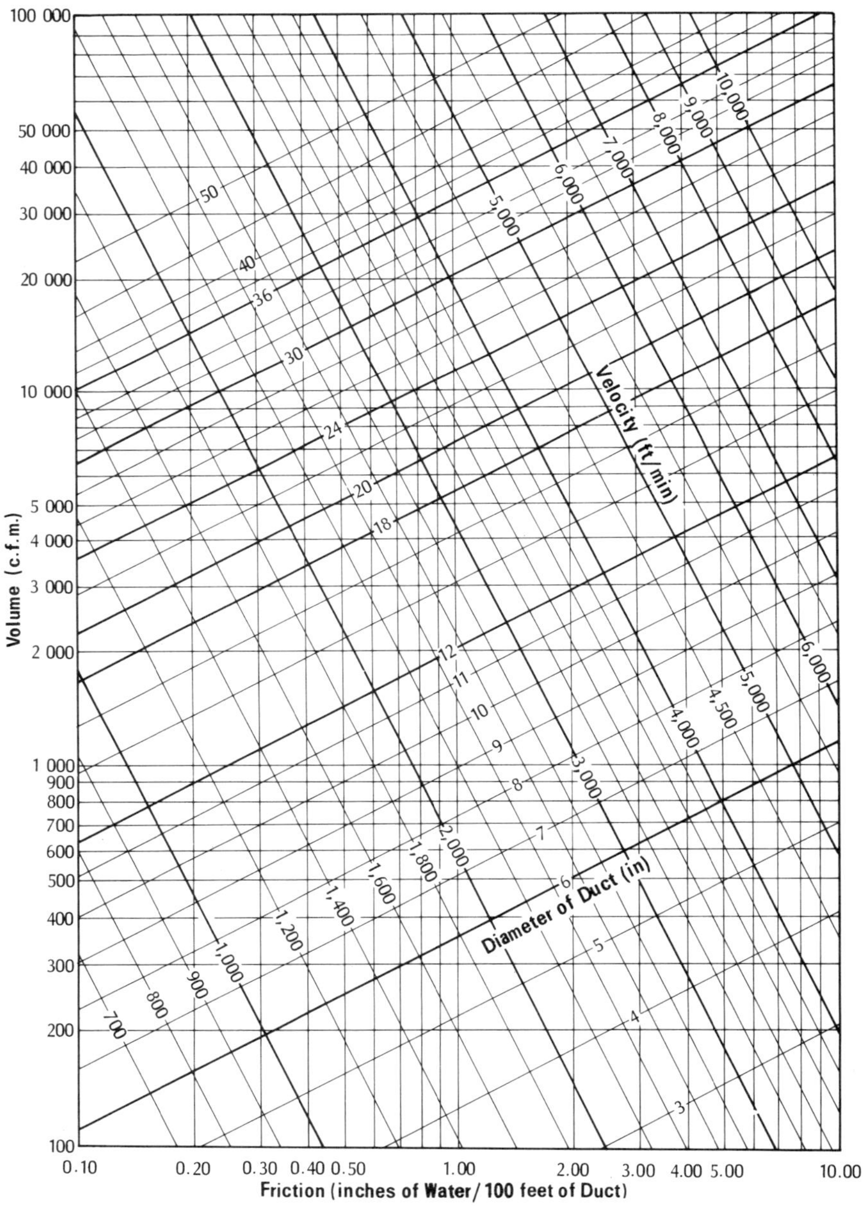

Figure 11.2. Friction losses in ventilation ducts.[4]

where the last two terms on the right-hand side represent the frictional losses due to various runs of duct along the system and the minor losses due to bends, hoods, screens etc., respectively, for a system of i duct runs and j minor losses. The actual variation of pressures along a duct may be seen in Figure 11.3, which uses the device of the energy grade line to visualize the variation in pressures. The diagram illustrates schematically the case of a hood, two ducts and an expansion upstream of a fan. The graph below plots the values of *TP* and *SP* and, hence, *VP,* since *VP* = *TP* - *SP.* The zero is atmospheric pressure so that upstream of the fan the total and static pressures are everywhere negative, and, since TP>SP, if the correct sign is considered, VP is always positive. From point 1 to point 2, TP drops due to frictional losses in the hood. Hence, the height labeled *a* in the figure is h_e for this hood. The velocity between 1 and 2 must increase from zero outside the hood to the value in the duct at 2. Thus the gap between TP and SP equal to VP must increase. The static value at the hood exit is SP_h. As it is difficult to define the average velocity at every point in the hood, the lines are shown dotted. From point 2 to point 3 the friction loss causes a further drop in total head, so that the loss in this run, h_f, is given by *b* in the diagram. TP and SP are parallel, since the velocity is constant for the constant area duct. From point 3 to point 4 the expansion causes a further loss in total head but a decrease in velocity head, since the duct area is now larger. The result is that at point 4 the static pressure has *risen.* (This important point should be noted. Total head in a system always decreases due to frictional losses except through fans and other prime movers. The static pressure, however, can rise or fall depending on the cross-sectional area of the duct system.) The run from point 4 to point 5 is similar to the run from 2 to 3, except that the loss per foot due to friction is less, since the velocity is less (see Equation 8). From point 5 to 6 a fan adds TP and, in the process, increases SP to a value slightly above atmospheric. Run 6 to 7 is another duct run with the exit SP ending at atmospheric as it must do. In the final section from point 7 to 8 the velocity head at exit decays in the air from its exit value to zero. Since the exact mechanism is not specified the line is shown dotted.

While a diagram such as Figure 11.3 is useful in visualizing the process, it is not clear from the diagram which one flow rate will satisfy all equations and result in the balance system described. Usually the analysis of a system must be completed before this diagram can be drawn in quantitative detail.

In addition to the friction losses and minor losses in a ventilation system, the *coefficient of entry* is frequently required. It is equivalent to a flow coefficient in fluid flow terminology. The *coefficient of entry,* C_e,

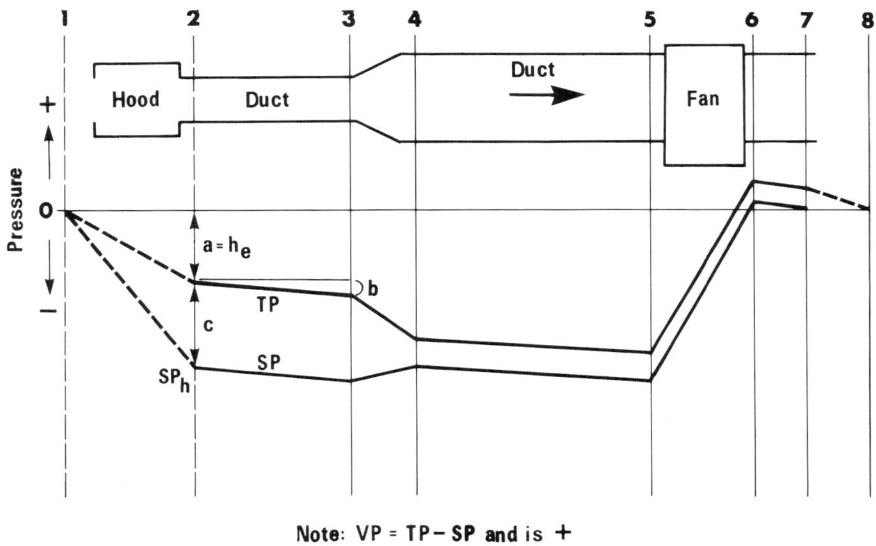

Figure 11.3. Variation of total static, and velocity pressures in a length of duct (schematic only).

is the ratio of the actual flow through a ventilation hood compared to the flow which would result if the entire static pressure drop across the hood had been converted to velocity head (or kinetic energy) without loss. The actual flow in a hood is:

$$Q = AV = 4005\ A\ \sqrt{VP}$$

The velocity pressure increases, if there were no losses, would be equal to the static pressure decrease and, hence:

$$Q\ \text{theoretical} = 4005\ A\ \sqrt{SP_h}$$

where SP_h is the static pressure drop across the hood.
Hence

$$\frac{Q}{Q\ \text{theoretical}} = C_e = \sqrt{\frac{VP}{SP_h}} \qquad (12)$$

and

$$Q = 4005\ A\ C_e\ \sqrt{SP_h} \qquad (13)$$

As with values of K, values of entry coefficient are available[2] for common hood shapes. Typical values run from 0.7 to 0.98. Equation 13 is useful in calculating the static pressure drop, SP_h, required across a given hood to produce a specified flow. The physical meaning of coefficient of entry may also be seen from Figure 11.3. The magnitude of VP across the hood is *c*. The value of VP if *a* was zero (no loss case) would be a + c or SP. The coefficient of entry on the diagram is then:

$$C_e = \frac{VP}{SP_h} = \frac{c}{a+c}$$

and if a = 0, C_e = 1.0 (ideal flow).
The coefficient of entry is also related directly to K since

$$h_e = K.VP \quad \text{from } 10$$

and

$$h_e = SP_h - VP \quad \text{from } 3$$

Therefore, equating and dividing by VP gives:

$$K = \frac{SP_h}{VP} - 1$$

or using Figure 11.2:

$$K = \frac{1}{C_e^{\,2}} - 1 = \frac{1-C_e^{\,2}}{C_e^{\,2}} \qquad (14)$$

Equation 14 enables values of C_e to be calculated if values of K are available. This is the usual case if data is available from nonventilation manuals.

Capture Velocities

The capture velocity of a particle is the velocity required to achieve a high probability of collecting a particle after being discharged into the air. In practice, the capture velocity does not depend on the particle fall velocity. This is because the particles of interest around 10-μm diameter sphere falls approximately 1.5 in./min at a density similar to that of asbestos. The capture velocity depends instead on the process of generation of the dust. As a result we shall attempt to record those recommended capture velocities available in the section entitled *Hoods and Enclosures,* p. 119. In general, however, capture velocity recommendations run between 50 to 100 fpm for enclosed conveyors through 400 fpm in certain types of grinding. Recommendations can go to 12,000 fpm for grinding, sawing and similar operations in which a very localized suction is used. In application in which the area over which the velocity is measured varies, such as the air intake on a portable saw or drill, recommendations are usually given as volume flow rates other than velocities.

The particular velocity recommended in any particular application depends on room air currents, movement of machinery in the vicinity and the velocity with which the dust is ejected from the process. These variables cannot be quantified in all cases so a fairly liberal safety allowance is required in practice.

In addition to capture velocities, minimum conveying velocities are also required. These are usually higher than capture velocities and are the minimum velocities required in ducts to prevent fallout and blockage. These values range from 3,000 to 5,000 fpm for asbestos with 4,000 fpm being the most common value. This is discussed further in the section entitled *Hoods and Enclosures* (see below).

DILUTION VENTILATION

Dilution cannot be recommended by itself as a solution to asbestos dust control. It is forbidden as a sole solution in many jurisdictions. It can, however, be used in conjunction with local ventilation. In addition, dilution ventilation is not as desirable as it once was, because the large quantities of air required must frequently be heated, which alters the economic balance toward local ventilation. Dilution calculations also require information on the rate of creation of dust and this is often impossible to obtain. In spite of these difficulties, it is obvious that high rates of replacement of air in a plant are more satisfactory than lower ones. In addition, the basic principle of arranging air flows to generally move from the operator toward the source of dust should be employed wherever possible. Air replacement rates of up to six air changes per hour have been used. The use of dilution ventilation must not be confused with the provision of makeup air for local ventilation. In this case, air is being provided to the work area to supply the design flows for the various local exhausts. The provision of makeup air is discussed in the section entitled *Analysis and Design of Systems,* p. 155.

HOODS AND ENCLOSURES

Enclosures or "booth"-type devices attempt to contain all dust as it is generated and are designed to prevent the escape of dust into the workplace. They are used for such operations as feeding, bagging or other work where dust is generated over a significant area.

Capture hoods attempt to entrain the dust from a localized source into a duct system. These can be divided approximately into high-velocity low-volume (HVLV) and low-velocity high-volume types. The first is used in places such as sawing and other portable machine operations and has velocities of the order of 10,000-12,000 fpm. The second is used in such locations as lathe cutting and boring.

In both types, the openings should be no larger than necessary, since excess entry area increases the volume of air that must be handled. Enclosure type hoods are not sealed; instead, specified air inflows are created for two purposes. First, the suction created by the fan to pull air into the enclosure ensures that the enclosure operates continuously at less than atmospheric pressures. As a result, all leaks in the enclosure, whether planned or not, will have an inflow that prevents dust escape. Second, a flow of air through the enclosure is required to convey the dust away and prevent buildup and blockage of the dust. Figure 11.4[7] shows a small enclosure used to protect an impact drill. The dimensions of the hood are approximately 12 in.^{2} The upper surface, where the drill rod enters, has a rubber seal to minimize the opening at the top. The sides, however, must be open to some extent to allow small rocks to escape. This necessity to allow some open area is typical of many operations. This particular system is self-contained and has the ducting and filter system mounted with the drill rig. A single value of capture or face velocity cannot easily be defined, since this will depend on the surface on which the hood is mounted and wear on the skirting. However, the air flow is specified at 1,500 cfm and this will give a face velocity around 3,000-4,000 fpm for a good mounting.

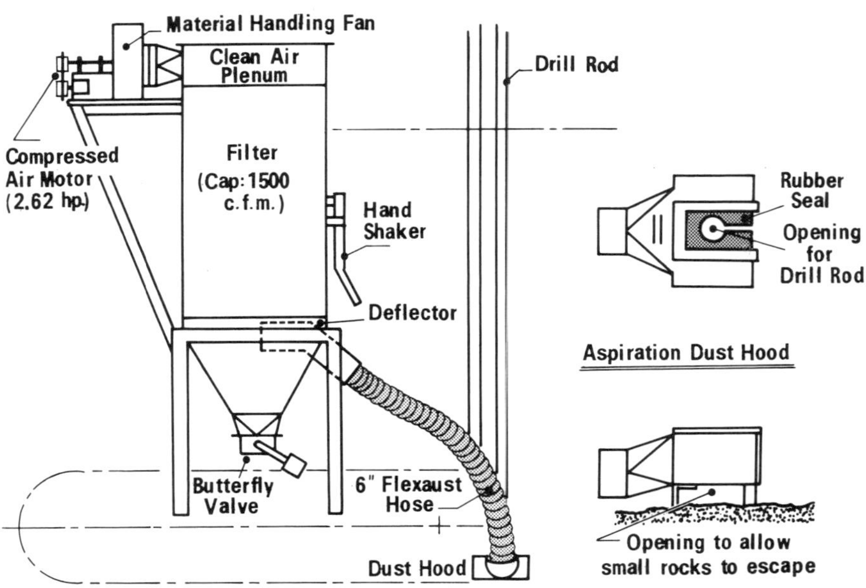

Figure 11.4. Dust control for an impact drill.[7]

Figure 11.5[7] shows a similar enclosure for a rotary drill, which is also mounted on the drill rig. Again, the volume flow rate (3,000 cfm) is specified, but the face velocity around the skirt depends on the way the skirt shapes itself to the surface.

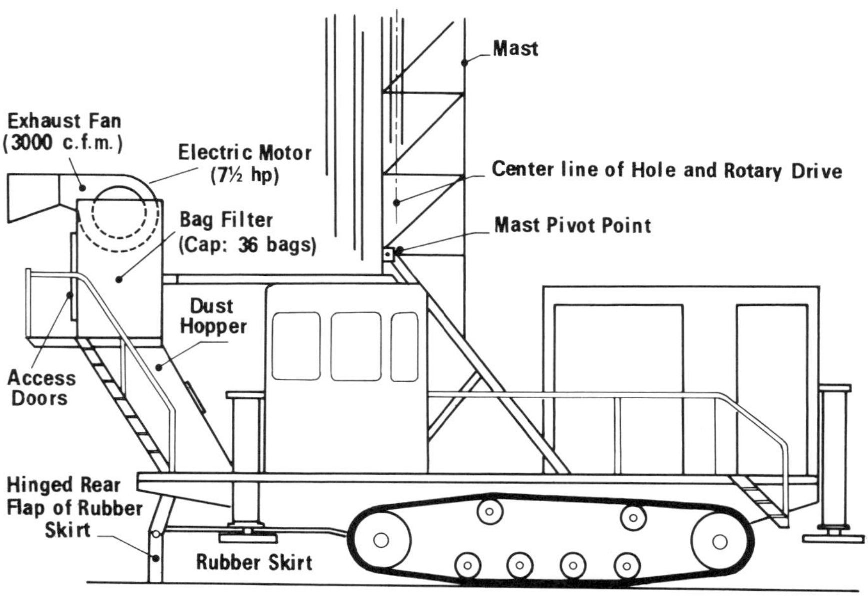

Figure 11.5. Dust control for a rotary drill.[7]

Conveyor belts are basic process elements in mines and textile plants. The conveyors can carry ore or fiber. There are several mechanisms by which dust can be generated on conveyor belts. The material has a velocity relative to the still air surrounding it, which can create dust by erosion. Loading and unloading of conveyors causes materials to impact in a way to create dust. The return belt can be a source of dust, particularly at the rollers. As a result, conveyor belts should be enclosed at all points and exhaust provided at the main dust emission points. Figure 11.6 shows a typical arrangement in a mill where ore is being conveyed.[7] The dust extraction occurs after the loading operation, at the base of the crusher, after the transfer point and from the bin at the dumping point.

The design of conveyor belt enclosures has been described in some detail in an excellent paper by Goldfield and Brandt.[8] Much of the material given here is based upon this work. A recommended design of

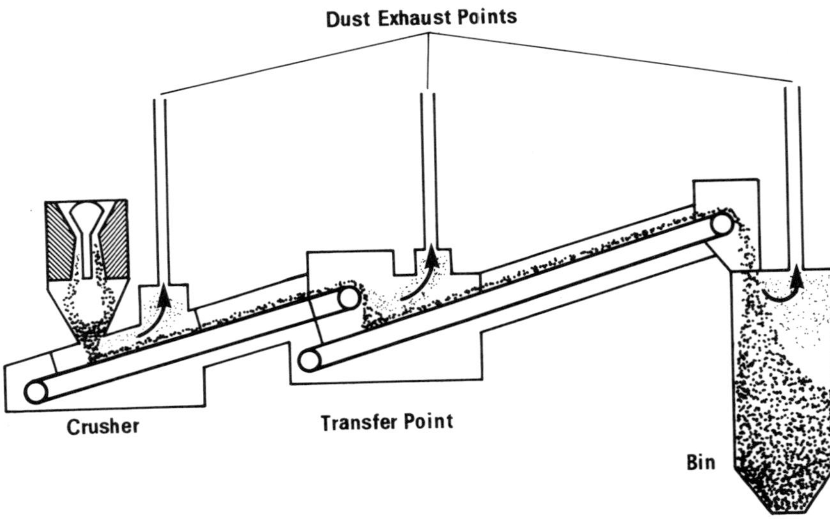

Figure 11.6. Schematic representation of dust extraction from a conveyor belt system transporting ore in a mill.[7]

conveyor enclosure at the loading point is shown in Figure 11.7. The particular arrangement is for ore and a rock box is located on the loading chute to break the fall of larger pieces. This would not be present for fiber due to the possibility of blockage. The ore enters along the line of the conveyor and at a smooth angle to it to minimize the impact on the belt. The exhaust duct is mounted on a settling box, which is present to enable the maximum amount of product to settle back on the belt without being entrained. An internal dust shield minimizes, but does not prevent, dust fall onto the bottom of the enclosure. Allowance must be made for air leakage into the enclosure sufficient to account for windage and to allow a face velocity into the settling box of around 150-200 fpm. Flows are commonly set at 200 cfm/ft^2 of open areas in enclosures of this type although some particularly tight enclosures may go to 100 cfm/ft^2 of open area. In all cases, however, the recommended conveying velocities in exit ducts must be met to prevent blockage and the entire enclosure should be kept at less than atmospheric pressure. Velocities into larger openings from work areas should be at least 200 fpm at all points. The loss coefficient, K, for the settling box can be taken as 0.5 approximately.

Figure 11.8 adopted from the same reference shows a form of recommended enclosure for conveyor belt dumping or transfer. The suggested velocities, as before, depend on the material to be conveyed. It is recommended that the scraper conveyor shown be used if necessary for ore-conveying applications[9]. For fiber-handling conveyors, a cleaning belt made

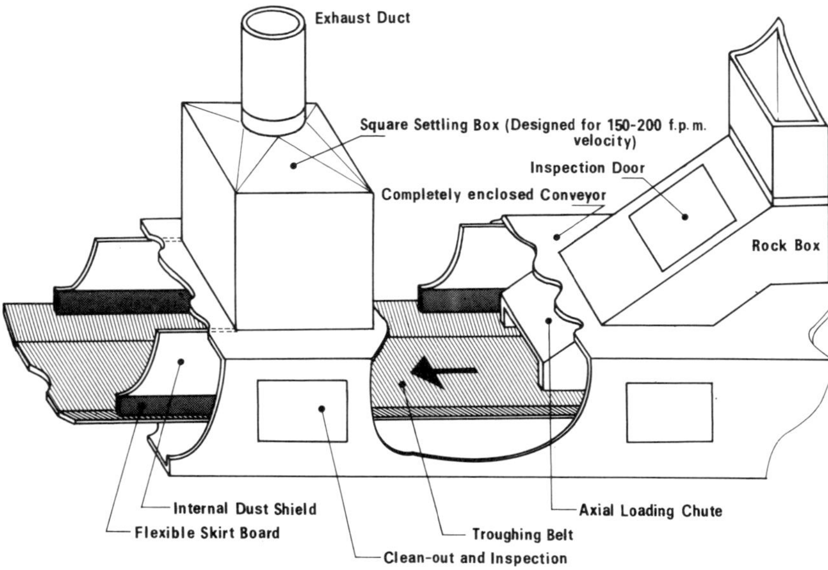

Figure 11.7. Conveyor belt enclosure in the loading region. This arrangement is intended for ore loading, and if fiber were being loaded, the rock box would be removed.

from conveyor belting with attached chevrons may be used. In both cases, the return belt scraper or brush should always be used to clean the conveyor.

Figure 11.9[8] shows an enclosure for the top of a vibrating screen. The flexible duct connectors can be made of nylon. Recommended velocities and flow rates vary, but recommended values lie in the ranges:

Flow: 200 cfm/ft^2 through hood openings, and
25 cfm/ft^2 of screen, where both are minima
Velocities: 4,000-4,500 fpm in duct
200-300 fpm in flexible connections

A recommended enclosure for a bag opening station for loose-packed bags is illustrated in Figure 11.10. Bags are placed in the enclosure by hand or they may be fed in from the right on a roller-conveying system (not shown in the diagram). The emptied bags are moved to the left into a shredder chamber or a plastic bag connected by an elbow to the opening chamber. If a shredder is used, it should be exhausted. The hopper should be exhausted to provide 100 cfm/ft^2 downward flow through the bottom opening of the hopper. The takeoff would be after the hopper. The

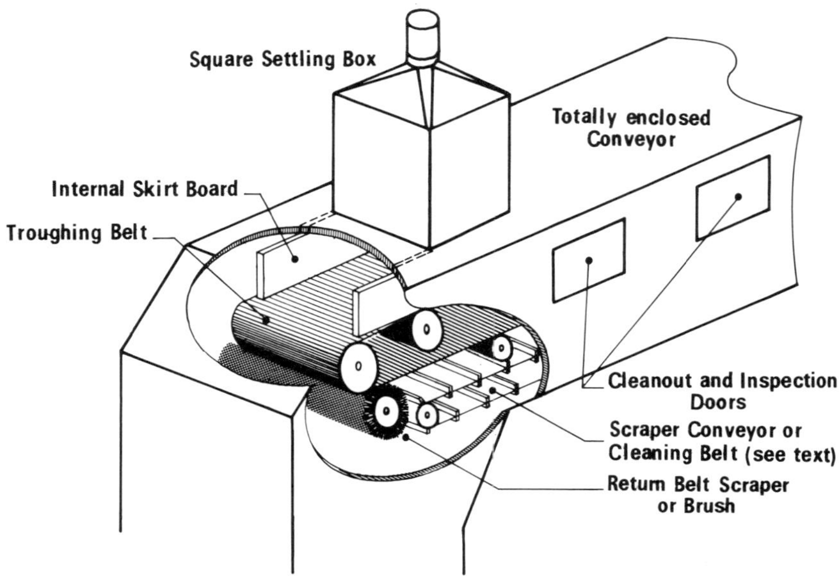

Figure 11.8. Conveyor enclosure for conveyor belt dumping or transfer.

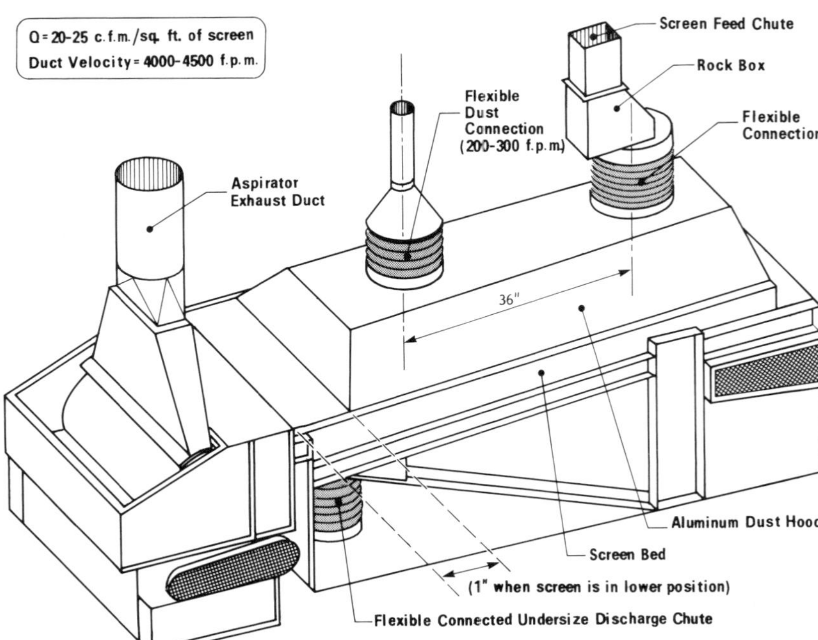

Figure 11.9. Gyratory fiber screen and rotary aspirator.

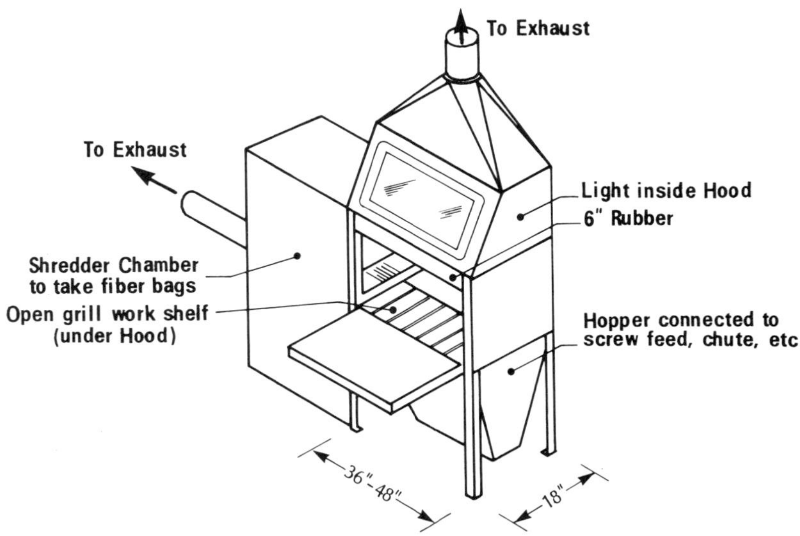

Figure 11.10. Bag opening station for loosely packed bags.

minimum velocity through the open area at front should be 200-250 cfm/ft^2 and duct velocities should be 4,000 fpm. Commercial hooding and equipment is available for pressure-packed bags.

Enclosures may be used for machining operations also. Figures 11.11 and 11.12 show two lathe enclosures recommended for asbestos pipe machining. Recommended volume flow rates are not given; however, the velocity in the exhaust ducting should be about 4,500 fpm if larger chips are present, and 300 cfm/ft is recommended in the *Industrial Ventilation* manual[2] for hood openings on lathes.

Figure 11.13 is an example of a high-velocity/low-volume exhaust[6] used for a table saw. Velocities around openings are very high, of the order of 12,000 fpm, but the degree of opening is very small, so that volume flow rates range from 10-250 cfm. The pressure drop required to drive this system is very high (of the order of 5 in. of mercury or 70 in. of water) and a high pressure drop centrifugal exhauster must be used. The arrangement shown collects dust from the top of the sawblade. As a result, the underside of the blade would need to be enclosed and a hopper for chips provided at the bottom. Alternative saw hoods are given in the *Industrial Ventilation* manual.[2]

The asbestos textile manufacturing process is traditionally a source of troublesome dust, and several enclosures have been described in the

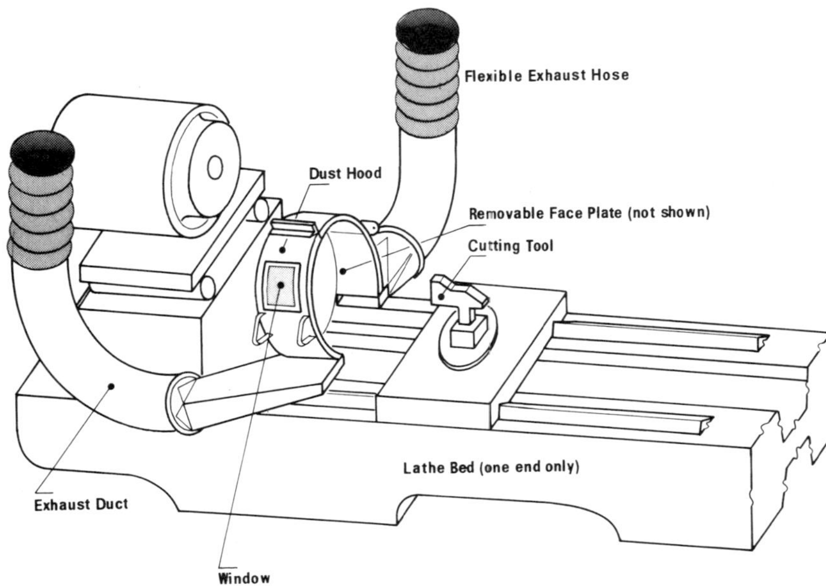

Figure 11.11. Pipe lathe dust hood.[8]

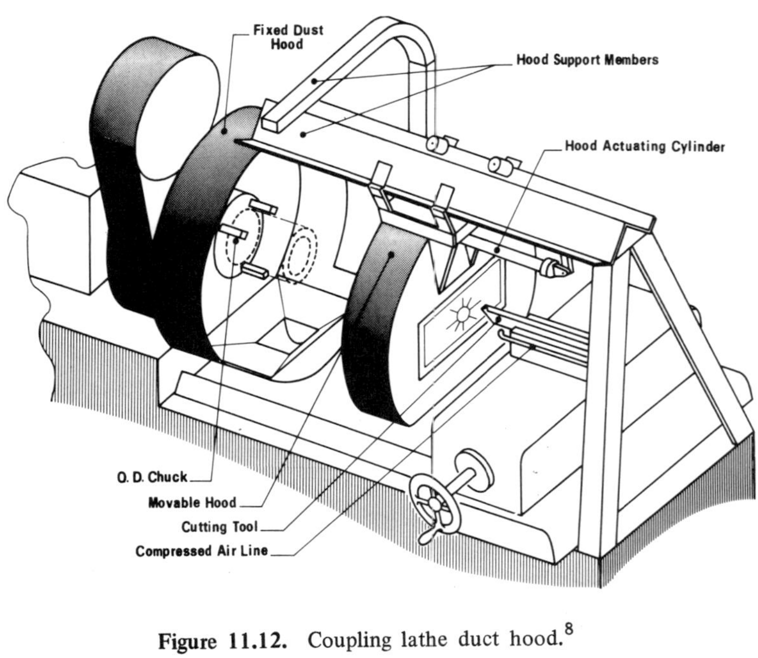

Figure 11.12. Coupling lathe duct hood.[8]

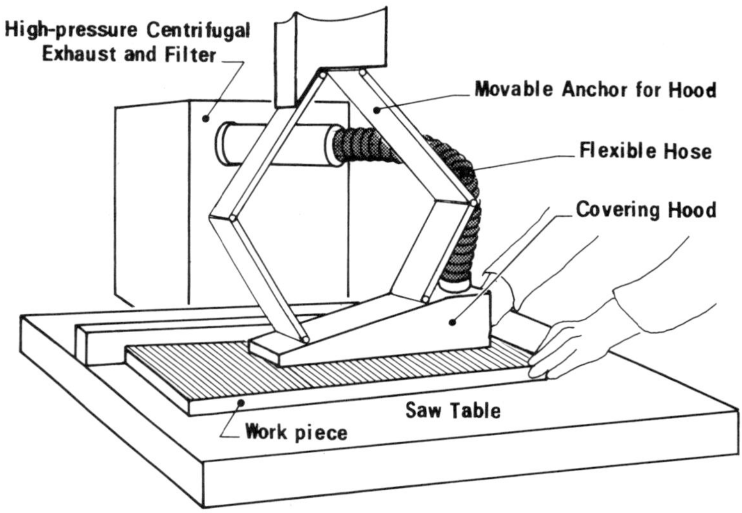

Figure 11.13. High-velocity enclosure for table saw.[6]

literature for the various processes. Goldfield and Brandt[8] have described a design of a feed box for a textile card machine. This design is shown in Figure 11.14. Velocities of ducts should be around 4,000 fpm, and flows of 200 cfm/ft should exist at openings during fiber-feeding operations. The entry loss has been given as K = 0.5, with the door closed and as the standard value for a tapered hood with the door open. This design provides for the feed of organic lap in roll form and, thus, eliminates feeding of blends to the card machine.

The textile fiber is often conveyed pneumatically in textile plants. At the end of a conveying process the fiber must be removed from the air. In mill practice this is done with a cyclone. Bamblin,[10] however, describes a drop box which is claimed to use less space than an equivalent cyclone for this purpose in textile plants. The device is illustrated in Figure 11.15.

Many other types of enclosures may be designed for particular processes. Those illustrated have attempted to give, by example, the usual arrangements found in good design. Other examples may be adapted from general ventilation practice.[2,3] Care must be taken, however, to keep velocities in a range suitable for asbestos dust. This often means that recommended values for general dusts are too low for asbestos control and those values mentioned should be adopted to the application.

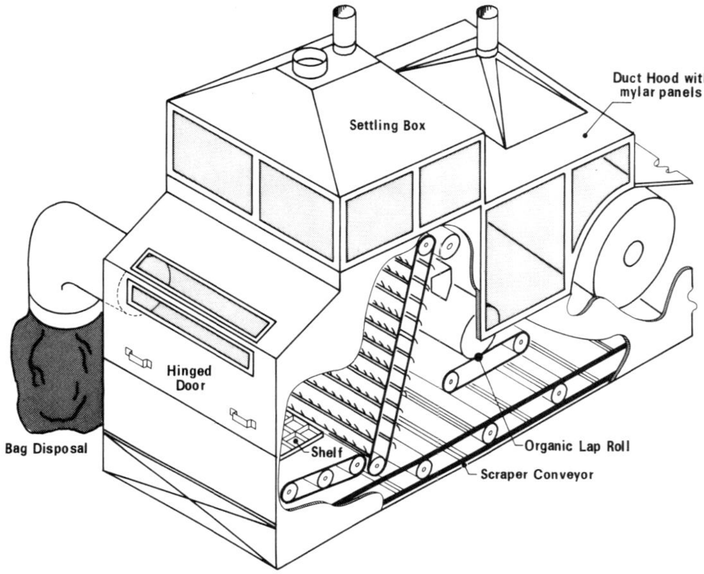

Figure 11.14. Feed box for textile carding machine.

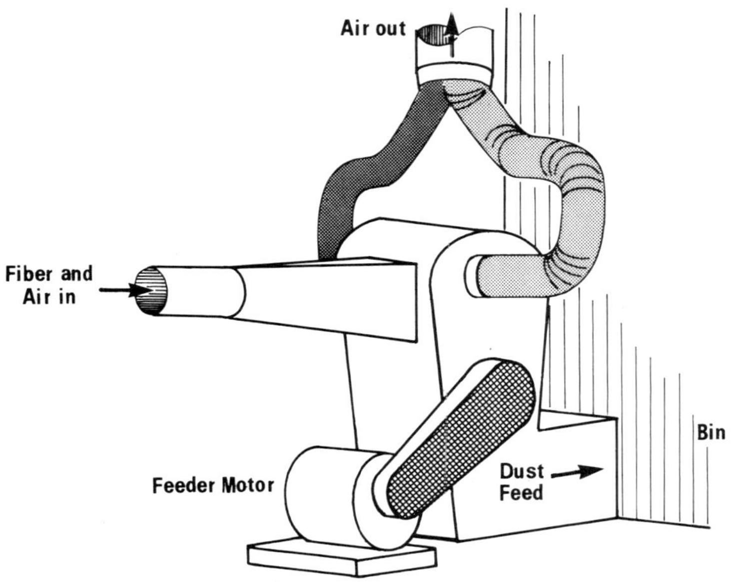

Figure 11.15. Drop box for removal of fiber from pneumatic conveying duct.[10]

Hoods differ only in degree of open area from enclosures and the difference is often only one of common terminology. A hood arrangement for a jaw crusher is shown in Figure 11.16 as described by Hutcheson.[7] Recommendations for this sytem are that duct velocities be 4,000-4,500 fpm and 150 cfm/ft^2 for open areas. Since the complete jaw crusher is not surrounded, the system would be called a hood, even though the entire feed system is enclosed.

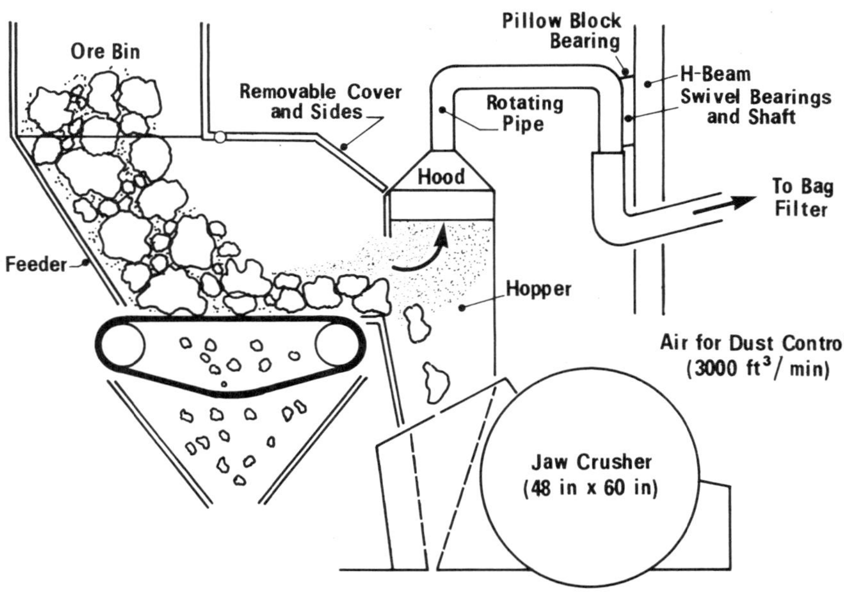

Figure 11.16. Dust control hooding for a Jaw Crusher Feed System.

A radial arm saw has a cutting blade, which moves over a large distance. One solution to this problem is shown in Figure 11.17, in which two collecting hoods are used. Duct velocities should be 4,000 fpm for normal saw speeds. The blade hood should come forward on the blade as far as possible consistent with the thicknesses being cut. An example of this is shown in Figure 11.18

A hood for a textile loom[11] is shown in Figure 11.19. In this case, the operator must have easy access to all parts of the operation. This has been solved by arranging hinged transparent panels at several points along the hood. With this arrangement there is good visibility during running when the panels are in place, and, if repairs or adjustments are necessary, the panels are easily moved away. This type of arrangement is easiest with

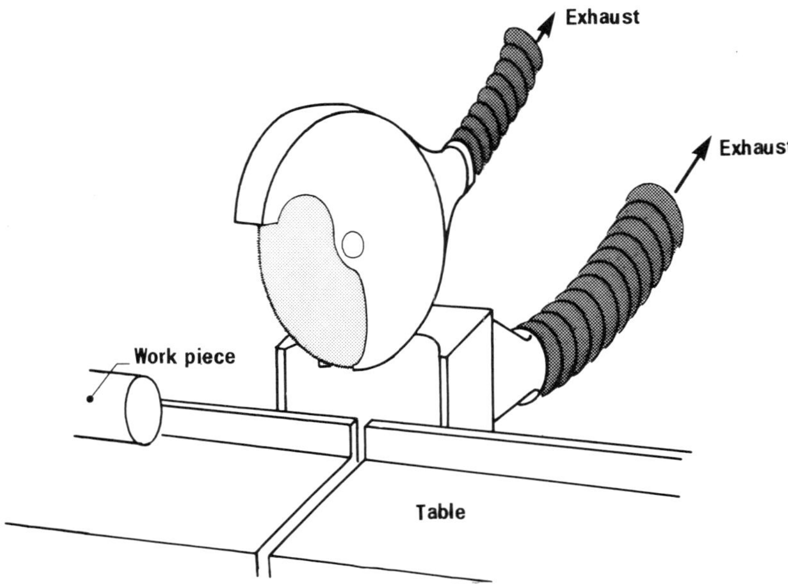

Figure 11.17. Dust hoods mounted on radial arm saw.

Figure 11.18. Highly enclosed saw blade on a radial arm saw. The operation is also continuously wetted.

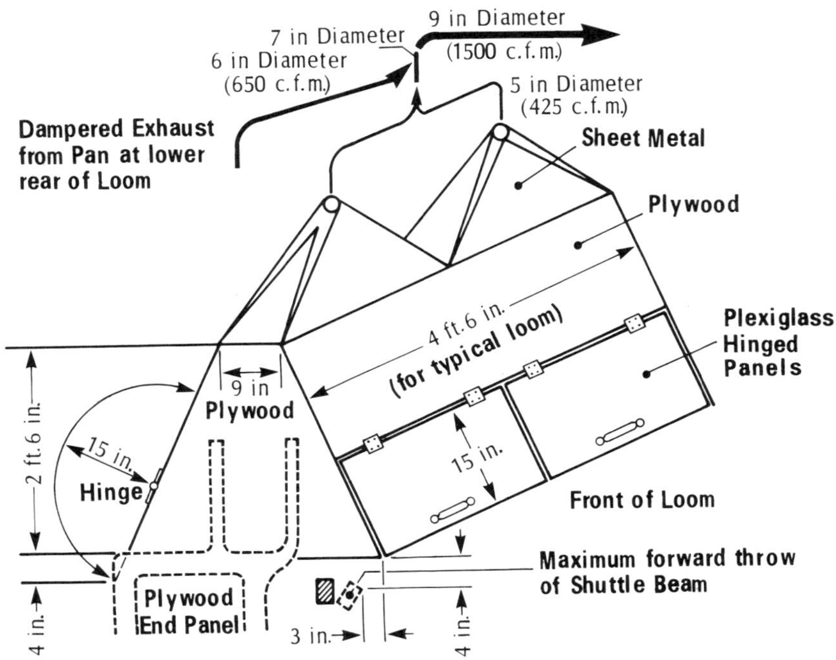

Figure 11.19. Hood for weaving loom.[11]

open hoods of this nature since the pressure under the hood is only slightly less than atmospheric. For enclosures, negative pressures greater than 0.5-1 in. of water can make removable panels difficult to open. Door-sized areas are difficult to open if internal pressures are below 0.1 in. of water.

The hood of Figure 11.19 illustrates another variation of the settling box used in Figures 11.4, 11.7 and 11.8. In these cases, it is useful to provide an enclosed air space where the larger dust particles can settle other than attempting to pull them directly up into the exhaust system. As before, of course, a vented hopper must be employed underneath the hood. This is indicated on the diagram. The upward velocity in the hood varies from about 100 fpm at the bottom to 300 fpm at the top of the hood. Again, this is similar to other recommended values for settling chambers. Duct velocities in this installation are around 3,000 fpm which in our opinion is low; 4,000 fpm would give less chance of blockage.

Drilling is an operation which normally does not create a large amount of dust. That dust which is created comes at the start of drilling, from chip breakage or from breakthrough of the drill at the bottom of the workpiece. Downward drilling causes the least dust and upward or horizontal drilling with portable drills releases much more contaminant. A form of hood that can be used for vertical drilling is shown in Figure 11.20.[12] A flow rate of approximately 500 cfm is recommended.

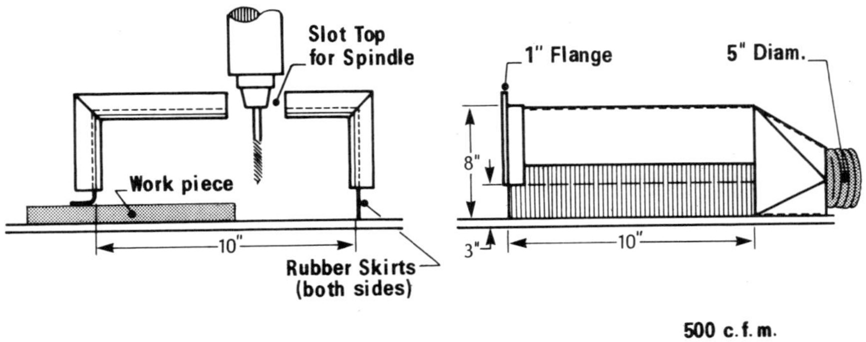

Figure 11.20. Dust control system for drill presses, hole saws, etc.

Portable Tools

Portable tools may also be hooded for dust control, and this possibility is particularly important in the construction industry. At present, commercial tools of several types are available in hooded configurations suitable for asbestos working. These include portable saws, drills, sanders and grinders. All fall within the "low-volume/high-velocity" (LVHV) category with velocities in openings designed to be over 10,000 fpm, although flows are ordinarily less than 500 cfm. Pressure drops in these systems are much higher than with hoods and enclosures and pressure drops of 8 in. of mercury are required by some systems. This type of flow can be provided by commercially available and portable centrifugal compressor units. An example of a system for sawing asbestos board is shown in Figure 11.21. The board is mounted on a prepared vacuum box with air extraction capability. Such a box can be constructed from wood if regular sawing of asbestos is to be done. If only occasional sawing is to be done, the work can be cut flat on a ½-in. plywood sheet. This prevents dust from entering the air on the underside of the board. Air is also extracted from the cutting region and the saw guard. Details of one configuration are shown in Figure 11.22. Note that the guard has additional shielding added to minimize the opening areas.

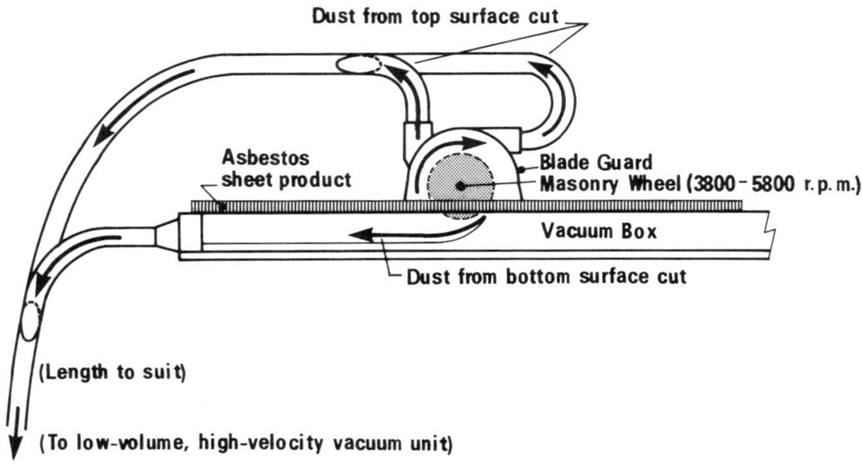

Figure 11.21. Portable saw dust control system.

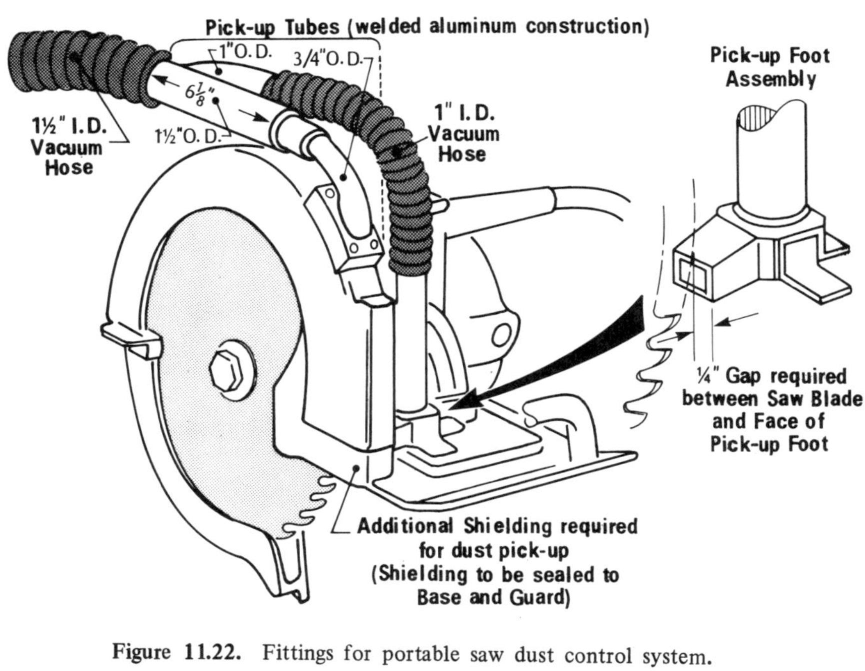

Figure 11.22. Fittings for portable saw dust control system.

Figure 11.23 shows two arrangements for portable drills. Figure 11.23(a) is a device that can be constructed if a considerable amount of vertical or horizontal drilling is to be performed. Figure 11.23(b) shows a commercial arrangement with a portable backpack vacuum supply. In this latter equipment the shroud on the drill sits flat on the workpiece and telescopes as the drill enters the work.

The various examples of local ventilation illustrated in this chapter are intended as examples of existing practices only. Other methods of control are always possible, and improvements in efficiency always pay in terms of both effectiveness and cost. Other recommended sources of effective designs beyond those already quoted are available in the literature.[13-16]

Recommended velocities have been the source of considerable discussion among experienced designers. This is particularly true of airflow and capture velocities. Here we have adopted the practice of quoting, without comment, those values which exist in the published literature. Table 11.1 summarizes these values for some of the more common applications and is mainly constructed from references already noted. When a range is given we would personally recommend the upper value if the following conditions are present:

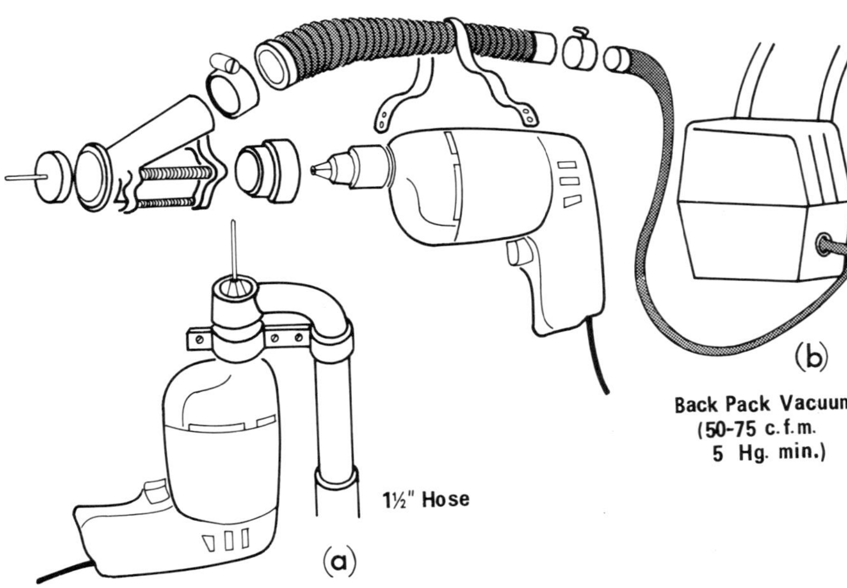

Figure 11.23 Dust control hoods for portable drills.

1. a possibility of larger than normal particles or chips entering the ducting;
2. sharp corners, dead air spaces, long horizontal runs or large numbers of dampers;
3. an insufficient number of openings and maintenance hatches provided to make cleaning straightforward;
4. a system not reguarly maintained and checked for blockages;
5. a system frequently stopped;
6. widely varying operating conditions such as moisture and work rate, opposite conditions of course implying that a lower velocity may be considered.

Table 11.2 illustrates typical dust fiber counts in the region of some standard handling operations. The recommendations of Table 11.1 are based upon these and similar measurements.

Table 11.1 Ranges of Velocities for Asbestos Dust Ventilation

Operation	Type of Hood	Air Flow or Capture Velocity	Minimum Transport Velocity (fpm)
Crushing	Enclosure	150 cfm/ft^2 of opening	4000-4500
Screening	Enclosure	200 cfm/ft^2 through hood openings but not less than 25 cfm/ft^2 screen areas	4000
Conveying	Complete	100 cfm/ft^2 of open areas	4000-4500
Bagging	Enclosure Booth	200-250 fpm through openings	4000
Mixing	Booth	200-250 fpm through openings	4000
Dumping	Booth	200-250 face velocity	4000
Mixing	Booth over the mixer	150-200 fpm face velocity	4000
Drilling of Panels Containing Asbestos	Moveable Hood	400 fpm capture velocity	Over 10,000 (low-flow, high-vel.)
Cutting of Asbestos Panels	Enclosure	600 fpm capture velocity	Over 10,000 (low-flow, high-vel.)
Bag Splitting	Enclosure	200-250 fpm face velocity	4000
Preform Press for Brake Shoes	Enclosure	200 fpm capture velocity	4000
Hot Press for Brake Shoes	Enclosure	200-250 fpm face velocity	4000
Grinding of Brake Shoes	Enclosure	400 fpm capture velocity at the tool rest	4000

Table 11.2. Measured Fiber Counts in the Region of Typical Local Ventilation Systems

Dumping	5 ft x 3 ft booth, face velocity 200 fpm	
	Area samples	0.20 f/cc
	Personal samples	0.62 f/cc
	To a mixer, no booth, face velocity at the mixer of 170-220 fpm	
	Area samples	1.2 f/cc
	Personal samples	0.62 f/cc
Mixing	3 ft x 2 ft booth over the mixer, face velocity 150-200 fpm	
	Area samples	0.24 f/cc
	Personal samples	1.4 f/cc
Drilling	Exhausted by a vacuum cleaner, capture velocity 400 fpm	
	Personal samples	0.2 f/cc
Cutting	Radial saw covered, exhausted 100 fpm at the dust generation point	
	Area samples	6.2-7.7 f/cc
	Skill saw exhausted under the asbestos panel, capture velocity under the panel 600 fpm, above the panel 100 fpm	
	Area samples	2.4-2.8 f/cc
Machining	Machine not exhausted	
	Personal samples	3.6-9.4 f/cc
Bag Splitting	On a conveyor under 1½ ft x 1 ft enclosure face velocity 200 fpm	
	Area samples	0.48 f/cc
	Personal samples	2.4 f/cc
Hot Press	Face velocity on the 4 ft x 2 ft opening is 240 fpm	
	Area samples	0.72 f/cc
	Personal samples	0.72 f/cc
Preform Press	Covered, and exhausted, 200 fpm at the dust generation point	
	Area samples	1.2 f/cc
	Personal samples	1.1 f/cc
Grinding	Face grinder, exhausted, capture velocity at the toolrest about 400 fpm	
	Area samples	0.96 f/cc
	Personal samples	6.2 f/cc

REFERENCES

1. American Society of Heating, Refrigerating and Air Conditioning Engineers. *Handbook of Fundamentals,* ASHRAE, New York (1964).
2. Committee on Industrial Ventilation. *Industrial Ventilation* - A Manual of Recommended Practice, 14th ed., American Conference of Governmental Industrial Hygienists, Lansing, MI (1975).

3. McDermott, H.J. *Handbook of Ventilation for Contaminant Control* (Ann Arbor, MI: Ann Arbor Science Publishers, Inc., 1976).
4. Danielson, J.A., Ed. *Air Pollution Engineering Manual,* U.S. Environmental Protection Agency, Research Triangle Park, NC (1973), p. 46.
5. Miller, D.S. *Internal Flow: A Guide to Losses in Pipe and Duct Systems,* British Hydromechanics Research Association, Cranfield, England (1971).
6. "The Control of Dust by Exhaust Ventilation when Working with Asbestos Dust," *Control and Safety Guide* #7, Asbestos Research Council, London (1973).
7. Hutcheson, J.R.M. "Environmental Control in the Asbestos Industry of Quebec," *Can. Inst. Min. Bull.*
8. Goldfield, J., and F.E. Brandt. "Dust Control Techniques in the Asbestos Industry," *Am. Ind. Hyg. Assoc. J.* 35:799 (1974).
9. Vaillancourt, R.M. Private communication, September, 1977.
10. Bamblin, W.P. "Dust Control in the Asbestos Textile Industry," *Ann. Occup. Hyg.* 2:54-74 (1959).
11. Postman, B.F. Dust Control in the Asbestos Textile Industry," *Am. Ind. Hyg. J.* 23(1):67-74 (1962).
12. "Asbestos in the Construction Trades," Construction Safety Association of Ontario, Toronto Ontario (January 1977).
13. NIOSH. "Recommended Industrial Ventilation Guidelines," HEW Publication #(NIOSH) 76-162, Supt. of Documents, Washington, DC (1976).
14. NIOSH. "Ventilation Requirements for Grinding, Buffing and Polishing Operations," HEW Publication #(NIOSH) 75-107, Supt. of Documents, Washington, DC (1975).
15. American National Standards Institute. "Fundamentals Governing the Design and Operation of Local Exhaust Systems," ANSI, New York (1971).
16. American Standards Association. "American Standard for Ventilation Control of Grinding, Polishing and Buffing Operations," ASA, New York (1966).

CHAPTER 12

AIR HANDLING AND CLEANING SYSTEMS

INTRODUCTION

Dusty air collected by the various devices described in Chapter 11 must be moved, cleaned and expelled from the work area. In this chapter, we shall outline the ducting, fans, dust collectors and other equipment necessary to process dusty air.

DUCTS

Ducts for handling asbestos-air mixtures are normally built of black or galvanized sheet steel. Local ventilation hoods are often connected to these ducts with flexible tubing which may be wire-wound or nylon. (For an application of nylon see the screening enclosure described on p. 124.) Goldfield and Brandt[1] recommended the sheet metal gauges shown in Table 12.1. These are slightly thicker than the gauges recommended for normal air handling applications.[2] The normal air conditioning recommendations may be seen in the third column of the table. The additional thickness is necessary due to the abrasive qualities of the dust mixture. Ducts are normally round where possible, because the velocity distribution in circular ducts does not have the regions of low velocity found in the corners of rectangular and square ducts.

Elbows for heavy-duty and abrasion resistance may be fabricated from welded 14-gauge sheet. Goldfield and Brandt[1] recommend a square design with removable back plate and rubber lining, if very high abrasion resistance is required. They also recommend that the square-round transition pieces and all fittings such as branch entries be rubber lined for these high-abrasion applications. The dust loading in these ducts runs from 0.2 to about 10 gr/ft^3.[3,4] Normal dust loadings from hoods are below 2 gr/ft^3.

Table 12.1. Sheet Metal Gauges for Ducts Handling Asbestos Fiber[1,2]

Diameter Straight Duct (in.)	Recommended Steel Duct for Asbestos (U.S. Standard Gauge)	Normal High Velocity Air Handling (U.S. Standard Gauge)
4-8	20	24
9-18	18	22
20-30	16	20
31-40	14	20
41 and over	14	18 or 16
High Abrasion Resistance	14 and 12	

In the design of dust handling ducts, considerable care must be taken to prevent arrangements in which blockage and high losses can occur. In fluid mechanical terms this means that separation must be prevented or minimized. A most important consideration in this regard is that flow must be moved "smoothly." Any change in velocity or direction must be done over as long a distance as possible. As a result, no duct branches should enter a main branch at 90°; 45° should be a maximum and $< 30°$ is preferable. These branches should never enter a main duct at the bottom to prevent dust fallout into the branch. The area of the main duct should increase when a branch enters it and only one duct should enter at a given point. Sudden area changes should be avoided both for enlargements and contractions. As a branch enters a main duct the size of the duct should be increased to about 10 or 20% greater area than the sum of the branches entering it. This will produce a system with the velocities decreasing along the run of the main duct. The number of bends and offsets should be minimized and bends should all be of large radius. Preferably, the radius of a bend should be at least twice the diameter of the duct.

All of these rules are variations on the basic attempt to prevent separation in the system. Some examples of good and bad practice are shown in Figures 12.1, 12.2 and 12.3, based on a series of guides in the literature.[5] Following these recommendations will minimize the possibility of plugging, blockages and high pressure losses. In addition, well arranged ducting is quieter than that which is poorly designed.

Dampers or blast gates are frequently installed at several points along the duct system. They are of little use in balancing systems and are often misused. We recommend that blast gates be installed only where necessary for control purposes. In our experience, untrained equipment operators can close blast gates if they are bothered by "drafts." In addition, gates

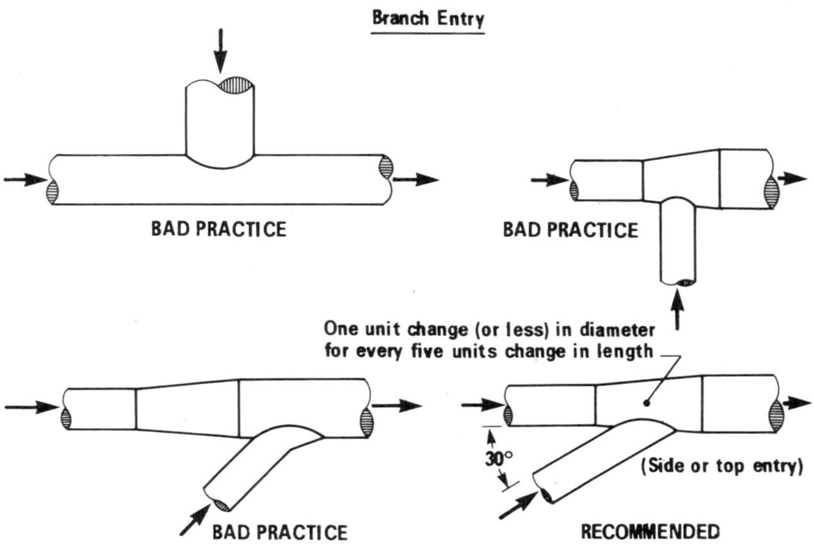

Figure 12.1. Possible geometries for branch entries in ducting.

can be inadvertently closed and not noticed for long periods of time. If left partly open they can be a fruitful source of blockage and plugging. However, blast gates are recommended where process equipment is not used continuously and where a particular duct can then be closed off. In some operations, blast gate controls are padlocked to prevent unauthorized adjustments.

In operation, all ducting must be inspected regularly. Plugging is often not obvious to equipment operators. In addition, partial or total plugging alters the air flow balance in the entire system. All ducting must be as airtight as possible, and joints should be taped or caulked. Holes worn in older ducting must be patched since these holes can considerably decrease the capacity of the system hoods. In older systems, corrosion or erosion of the fan wheel or fan belt slipping will reduce volumes and velocities. It is our experience that many properly designed collection systems can become ineffective due to lack of maintenance.

DUST COLLECTORS

Many types of dust collectors are used in cleaning asbestos dust from the air. The most effective and popular of these at present is the fabric

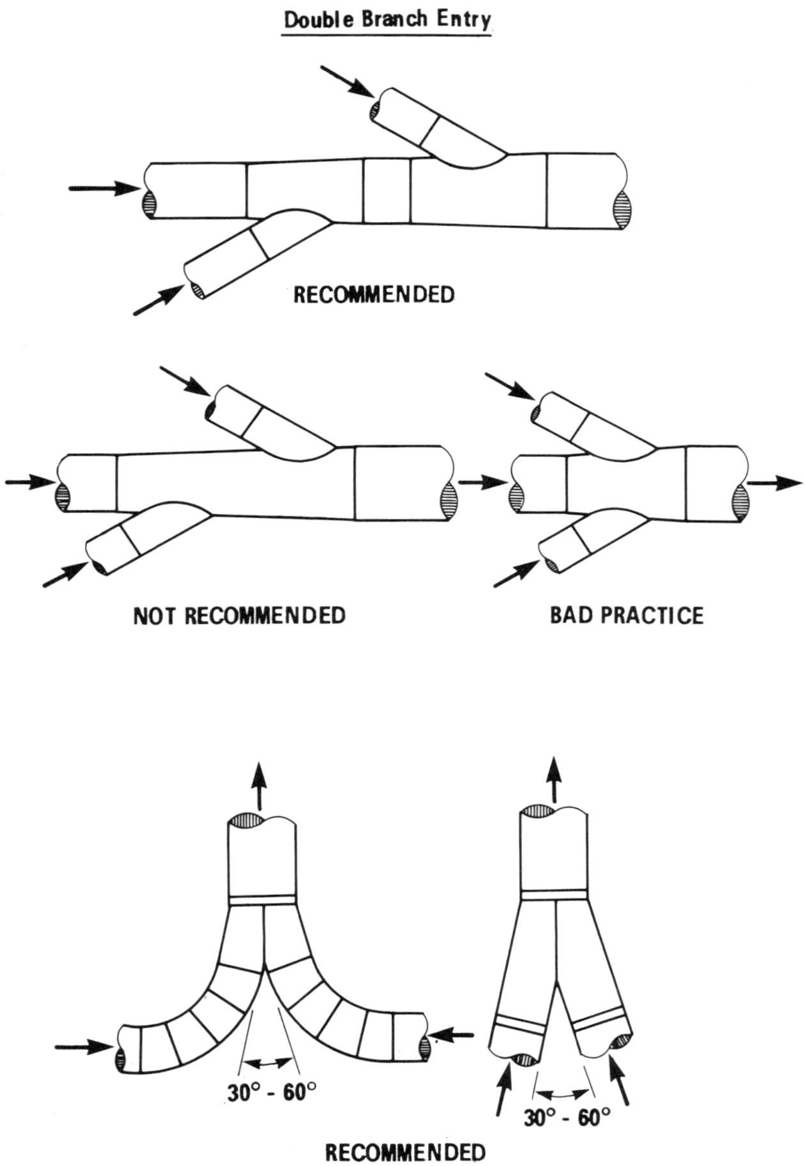

Figure 12.2. Possible geometries for double branch entries in ducting.

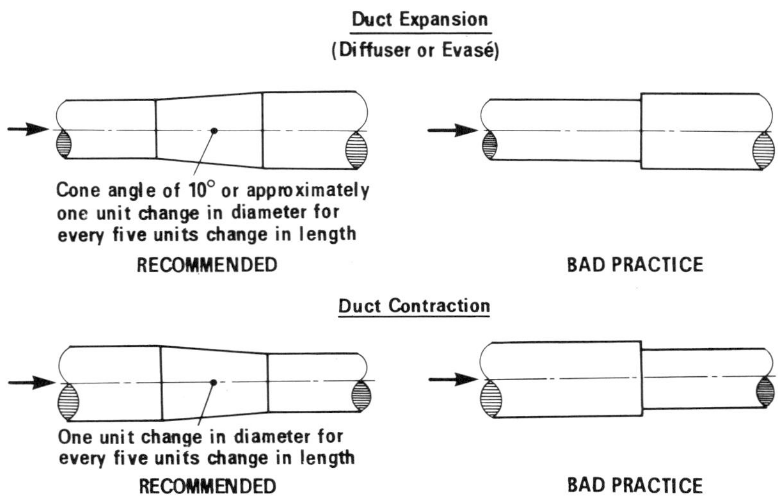

Figure 12.3. Expansions and contractions in ducting.

filter (or bag filter or baghouse). For process operations in which fiber must be removed from the air, cyclones are the most common device. Cyclones are not efficient enough for final air cleaning, yet are frequently used for precleaning dust from air that is later filtered.

Wet collectors in the form of low-energy scrubbers have been installed in asbestos operations;[1,6] however, they are not at present of high enough efficiency and suffer from plugging, freeze-up and corrosion.

Electrostatic precipitators have also been installed in asbestos cleaning operations but these have never been popular. The electrical properties of asbestos dust are not suitable for electrostatic precipitation in most cases. There have been some attempts to alter this by controlling the wetness of the dust and the humidity of the conveying air. At present, however, the process must be considered experimental for this type of application.

A basic bag filter of the most common industrial type is shown in Figure 12.4. It consists of a series of bags usually around 5-8 in. in diameter and anywhere between 5 and 15 ft in length. Dusty air enters a chamber beneath the bags and is drawn through the inner surface of the bags. As a result, clean air enters the chamber exterior to the bags and is drawn upwards. The dust collects on the inner surface of the bag. Periodically, the bag is shaken or a reverse pressure is applied. As a

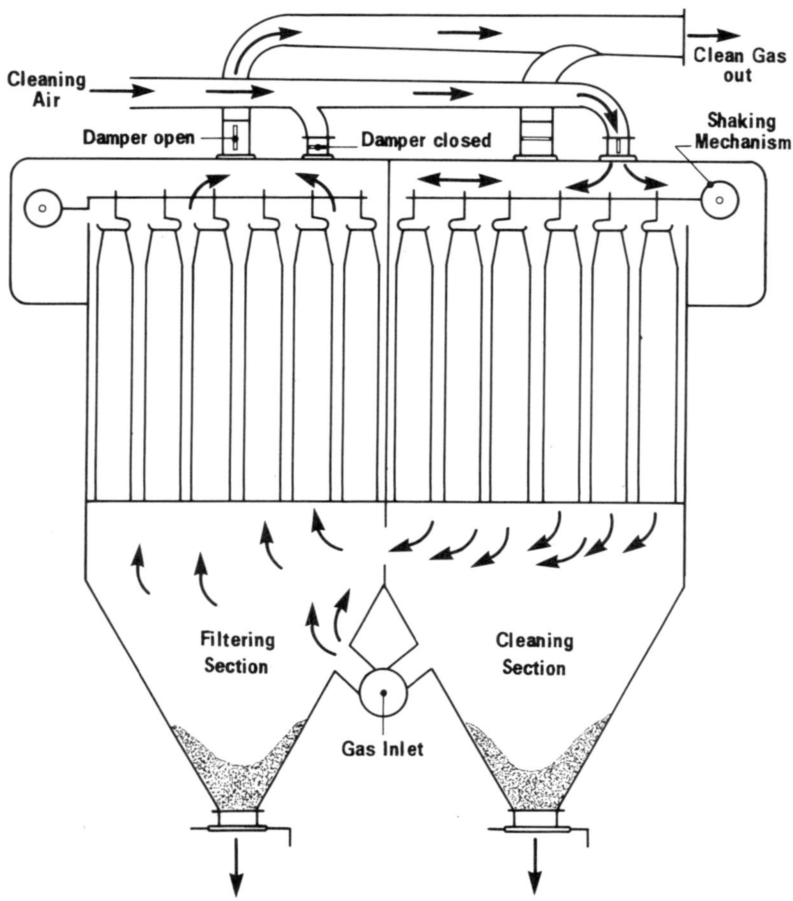

Figure 12.4. Schematic diagram of a basic bag filter. This filter employs both reverse flow and shaking to clean the bags.

result, the cake formed on the inner surface of the bag breaks up and is deposited in the hopper at the bottom of the chamber. Many variations of geometry flow direction and bag arrangement are possible. Cleaning procedures such as shaking, reverse pressuring and air jet cleaning are employed in different applications. In the particular filter shown, the right-hand section is being cleaned by both a shaker mechanism and by cleaning air. Normal asbestos practice uses shakers only.

In large filtering operations[4,6] it is normal to pretreat the air with cyclones and large dusty air plenums before feeding the air to the baghouse. The plenum serves to allow further fallout of dust, thus allowing less loading on the bags. and also serves in the larger systems as a source of constant negative pressure. As a result, additional ducting systems may be added by attaching them to this plenum. The arrangement is shown schematically in Figure 12.5.[6]

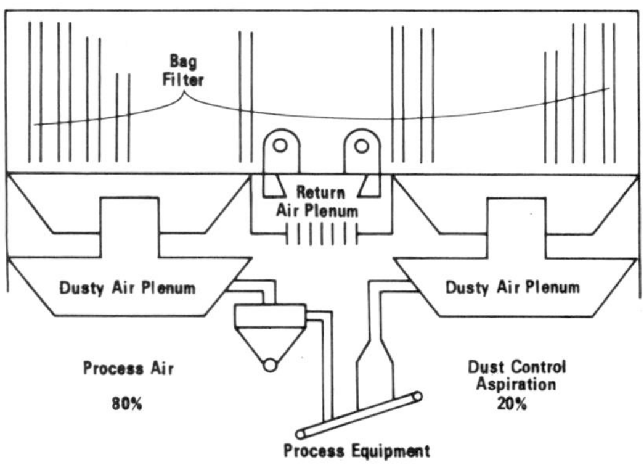

Figure 12.5. Air flow to the bag filters in a large asbestos mill.

The most important parameters in selection of a bag filter for a given application are the total amount of cloth surface to be used and the filter material. Cloth surface is ordinarily sized by a quantity called the *air to cloth ratio.* This is the ratio of air flow in cfm to the total cloth area in square feet. Hence, this is actually the face velocity in feet per minute through the cloth. Values for asbestos normally run around 2-3 cfm/ft^2. This value enables a rough estimate of filter size to be made very quickly.

For normal temperatures, cotton sateen is a recommended filter material. It is both efficient and long wearing. In one case, a life of six to seven years has been reported (Table 12.2).[4] Hutcheson[6] recommends outdoor bags of silicate-treated nylon acetate for outdoor applications. High temperatures require fiberglass, Teflon®* or similar heat-resistant materials. In another application, using a reversed pulse cleaning action, considerably higher air-to-cloth ratios were used.[7] In this case, an air-to-cloth ratio of about 8 fpm was employed. This is typical of this particular type of filter cleaning mechanism, which allows higher filtration velocities than normal baghouses. The cloth used was felted Dacron.®

*Registered trademark of E.I. duPont de Nemours, Inc., Wilmington, Delaware.

Table 12.2. Fabric Filter Applications–Asbestos Mining and Asbestos Manufacturing[4]

Application	Type of Collector	Type of Cloth	Length of Bags (in.)	Diameter of Bags (in.)	Air to Cloth Ratio	Expected Pressure Drop (in.)
Asbestos Milling	Continuous	Cotton Sateen	168	5	2.5-3.0	2.5-4
Asbestos Ore Dryers	Continuous	Orlon®	168	5	2.5	1.5-2.0
Asbestos-Cement Raw Material Handling	Continuous	Cotton Sateen	126	5	2.5	3.0
Asbestos-Cement Finishing Machines	Intermittent	Cotton Sateen	168	5	2.0	1.5-2.0
Textile Carding	Intermittent	Cotton	126-168	8	5	1.5-2.0

Typical pressure drops to be expected in fabric filters are given in Table 12.2 and range from 2-4 in. of water. Pressure drops in reverse jet bag houses are higher and range around 4-6 in. of water. This may be compared with a typical pressure drop from atmosphere to plenum in a large duct system of about 5-6 in. of water. Smaller systems with shorter runs of ducting will often have pressure drops to the filter inlet of 3-4 in. of water. As a result, in normal systems the filter accounts for approximately one-half or one-third the total pressure drop in the system.

During filtration, the buildup of dust on the filter cloth produces an increase in pressure drop with time if we assume constant volume flow in the system. Figure 12.6 shows what the pressure-time variation in a bag filter would look like for this case. In practice, real systems do not provide constant volume flow. This point will be discussed further in the section entitled *Fans and Compressors*, p. 152.

Hoppers and disposal devices at the bottom of bag filters and dusty air plenums require special consideration with asbestos dust. Most hoppers are under vacuum or occasionally a positive pressure depending on the fan location. In either case, they require valving at the bottom to let the collected dust out without admitting air to reentrain and redeposit dust or causing air flows through the discharge.

Hoppers must be designed to prevent blockage of the flow of the dust through the exit. Most hoppers have tapering sides to facilitate this flow. The usual angle to the horizontal is 60°, which is greater than the typical angle of internal friction for dry dust. However, hoppers can block if the dust is damp or the hoppers allowed to overfill. Under these circumstances, vibrators and mechanical rappers may be necessary.

Valving equipment at the hopper exit can be of several types. Automated equipment usually uses a rotary valve as shown in Figure 12.7a. For larger units rotary screw conveyors are used downstream from rotary valves (Figure 12.7b), and for noncontinuous operation, trip gates or discharge drums are used as in Figure 12.7c.

The efficiency of bag filters is extremely high. It is not unusual for a commercial installation to filter over 99.995% of the mass fed to it. The problem with asbestos dust is that the very small fibers must be collected independently, whether or not the larger particles have been picked up. Since the mass varies as the cube of the diameter for a sphere, it is easy to see that a mass-based efficiency is not a suitable guide to the cleanliness of the filtered air. Very little published material is available on the air quality at the outlet of the fans. Goldfield[4] has reported one case of a very large installation at a Canadian mill in which the air exhausting from the filter has a solids content of 183 $\mu g/m^3$ by gravimetric sampling and a fiber count of 0.5 f/cc using the membrane filter technique.

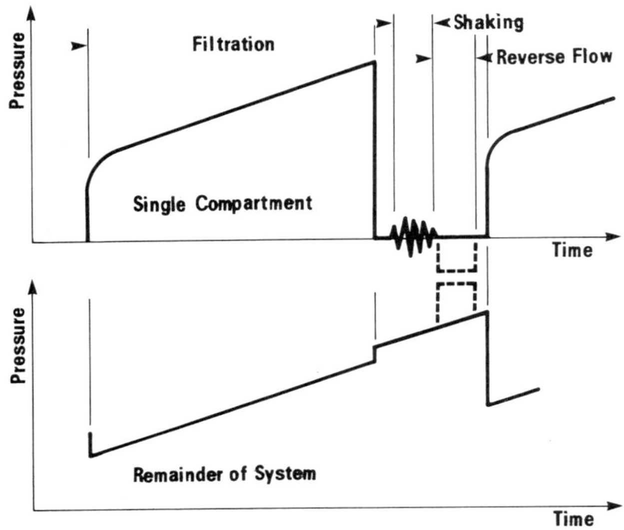

Figure 12.6. Pressure variation across a compartment during one cleaning cycle (assuming constant volume flow).

Bumbaco *et al.*[8] have reported on a large number of fiber counts from dryer stack samples. These were ordinarily below 0.5 f/cc and averaged around 0.2 f/cc. It is to be expected that any holes in bags would degrade these values temporarily, and Goldfield[4] reports rises in dust levels during cleaning. Synthetic cloths were said to be worse in this regard. These values can be compared to European standards for factory air of 100-150 $\mu g/m^3$ by gravimetric measurement and the 2 f/cc standard frequently used for fibers.

Air of the quality reported above is not considered suitable for recirculation.[5] In general, carcinogenic substances are not recommended for recirculation under any circumstances and the other contaminants are recommended for recirculation only if the air quality is no worse than 20% of the TLV. Asbestos, although a carcinogen, is an exception because of the ability to obtain high filtration quality.

In recent years, however, the tremendous cost of heated makeup air in winter has encouraged people to look into high-efficiency filters, through which filtered air may be returned to the workroom if the concentration of asbestos in the effluent air is quite low. The following criteria have been recommended:[9]

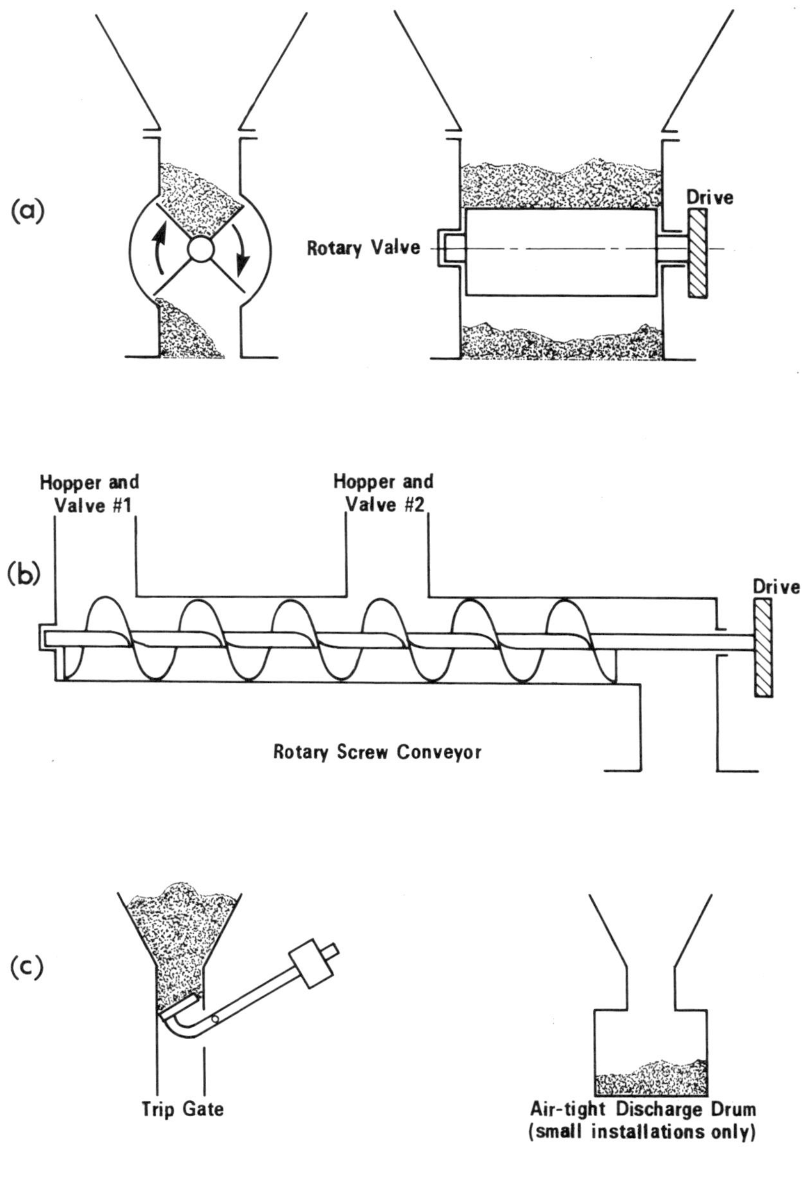

Figure 12.7. Hopper disposal mechanisms.

1. An acceptable monitoring device should be placed in the return air stream to continuously measure the dust concentration. The device should be fitted with an alarm system to sound when the dust count exceeds one-half the proposed TLV.
2. The monitoring device should be calibrated at least once a month when operating against a standard dust-counting technique.
3. The monitoring device should be checked at least once a day when operating.
4. A dust-counting survey should be carried out by the company semi-annually.
5. The collector system should be equipped with a damper system so that air can be directed outside during warm weather or when the alarm sounds.

The following is a set of requirements that has been proposed by Astleford[10] for the monitoring system.

1. The monitor should be online and automatic.
2. Sampling and analysis should be performed continuously with a sampling, analysis and display cycle not to exceed 15 minutes.
3. The monitor should be specific for a given contaminant, *i.e.,* interference free.
4. The minimum detectable concentration should be less than that which is predicted by the theoretical recirculation model.
5. There should be a digital display of concentration for comparison with a redundant manual sampling of the air cleaner effluent.
6. There should be an analog output proportional to concentration for input to a fail-safe control system.

Figure 12.8 shows an example of recirculation[5] using an ultrahigh-efficiency filter. These High-Efficiency Particulate Aerosol (HEPA) filters were originally designed to filter highly toxic agents, bacteria and radioactive particles. They are used when dealing with these materials, in filtering air for clean rooms and in other high-efficiency air cleaning applications. They are occasionally called absolute filters. For these high-efficiency circumstances, the term decontamination factor (DF) is used where

$$DF = 1/(1-n)$$

where n is the efficiency. Decontamination values of 10^6 can be required in special radioactive applications. Efficiencies of this magnitude can be achieved with deep beds of filter media. The subject of extremely high-efficiency air cleaning has been dealt with by White and Smith[11] and by Keilholtz and Battle.[12] Pressure drops of the order of 0.4-1.5 in. of water are typical for face velocities of 4-30 fpm. The efficiency of this type of filter is commonly measured using either an aerosol of dioctylphthalate (DOP) with mean size around 0.3 μm or methylene blue with a mean

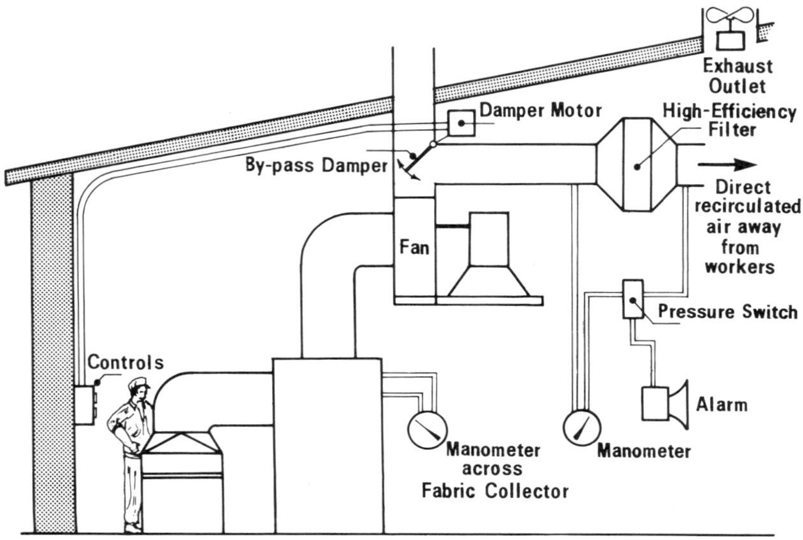

Figure 12.8. Example of a possible recirculation system for asbestos dust.

size around 0.5 μm. DOP is used in the U.S., while methylene blue is used in Britain.

The high efficiencies necessary for this type of filtration may be achieved by deep beds of collection material such a fiberglass, cotton or even asbestos, and mixtures. Typical bed depths for this type of filter run from 4-20 in. High efficiencies may also be obtained with papers made of asbestos, glass, etc. These papers are commonly arranged in corrugated or honeycomb fashion to maximize space utilization and may also be used in respirators.

Because of the very high efficiency of these filters, they are easily clogged by significant amounts of heavier dust particles. As a result, air must be prefiltered before going through a high-efficiency filter. These filters are extremely sensitive to damage by rough handling, high temperatures, moisture, etc. The quality of their output air is, of course, extremely sensitive to leakage or filter damage.

Absolute filters are not in common use in the asbestos industry at the present time. However, if dust standards become stricter and the cost of heating increases significantly, we can expect to see increased interest in this type of device.

FANS AND COMPRESSORS

In dust collection systems, fans may be installed either before or after the filter. Fans installed before the filter must, of course, be capable of running in the dusty atmosphere. Characteristics for the three most commonly used types of fans are shown in Figure 12.9. Forward-curved blade fans or volume fans are normally used for large volumes of air at low pressure. As a result, they can be used at lower speeds than the other types and, therefore, usually have relatively quiet operation. These fans are relatively cheap. Since the horsepower curve rises steeply with increases in volume, this fan is appropriate for use only in locations where the resistance against the fan is fixed at the design value. This type of fan is not suitable for dust handling since dust will collect on the curved blades. These fans are used extensively in heating and ventilating and in air conditioning work, but are not normally used for air pollution control devices.

Backward-curved blade fans have a relatively smooth static pressure variation, with a peak near the zero volume point. An important aspect of these fans is that they are self-limiting. The horsepower required at constant rpm reaches a peak before the maximum volume is reached. As a result, running under low resistance conditions does not endanger the motor, and the constantly varying demands of a complex duct system will not result in overloading the motor if a fan of this nature is used. Ordinarily, fans of this type are large for a given application. This type of fan is very popular due to its self-limiting power requirement and is very commonly used to drive dust-handling systems. However, dust may not be sent through this fan because buildup will occur on the blades. As a result, this fan must be used in forced draft applications or downstream of the filter.

The radial blade fan is the fan type commonly used for dusty gas streams. Many special designs have been developed for specific dust handling and pneumatic conveying problems. Efficiencies and costs are relatively less than the other types. The design is not self-limiting on power.

Figure 12.10 illustrates the interaction of a fan and the system. ΔP_d is the resistance in the system of the ducting, dampers and other passive elements. In the vast majority of applications, the flow is turbulent, and ΔP_d is proportional to the square of the volume flow rate, Q^2. ΔP_d is independent of time and may be taken to include those losses in the filter other than that through the actual filter medium. Hence, exit, entry, turning losses, etc. can be included in ΔP_d. The loss due to the cloth itself is time-dependent, as illustrated in Figure 12.6 and is shown as $\Delta P(t)$ on the diagram. This pressure drop is linear with Q at an instant in time. The point where than fan characteristic intersects the sum of ΔP_d and

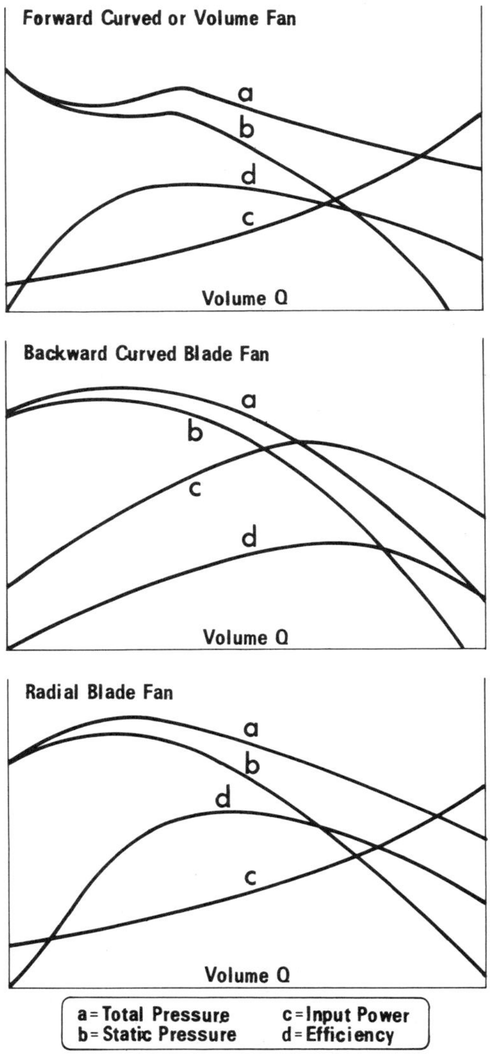

Figure 12.9. Fan characteristics for dust collection systems (constant rpm).

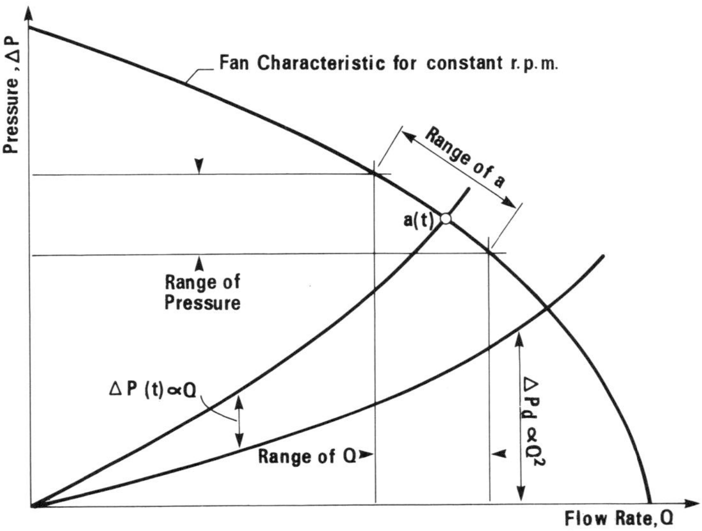

Figure 12.10. Interaction of duct system and fan (schematic).

ΔP(t) (total system resistance) is the operating point of the system. As a result, point a(t) represents the normal operating point of the system. The resulting operating range of ΔP and Q are indicated schematically on the diagram. We can see that neither a constant pressure nor a constant volume flow rate model is exactly appropriate for real systems. A more accurate model may be achieved by analyzing the data in the form of Figure 12.10. For many systems, however, the variation in pressure and/or flow is relatively small. This is particularly true if system resistances other than the filter medium comprise the largest part of the system. In large or expensive systems, of course, a detailed analysis of all possible operating ranges is necessary.

Another type of instability whose existence is not generally realized can occur in dust-handling systems. If a velocity at a particular point is too low to keep dust in suspension or to reentrain dust that falls out during a shutdown of the system, then dust will accumulate at this point. This obstruction can cause an increase in velocity locally, and the dust will eventually reach an equilibrium state in this location and no further accumulation will occur. If, on the other hand, the dust blockage causes blockage in the system sufficient to *decrease* in the duct flow so much that a local increase in velocity does not occur with the change in area, then blockage will occur. Whether this will happen depends on a detailed analysis of system behavior around the operating point in Figure 12.10.

The other type of air-moving device used in dust control is the high-pressure low-volume device used for the small enclosures around saws, drills, etc. and for vacuum cleaning. These units are single- or multiple-stage centrifugal compressors. They are universally backward-bladed because of the wide range of loads to which they are subjected. This is true even of the small number of units with the compressor on the dirty side of the filter. These units are commercially available and are usually portable to some degree. They range in flow from 50-500 cfm and in pressures from 1-20 in. of mercury. Horsepower ranges from 1-10 in the most popular units.

ANALYSIS AND DESIGN OF SYSTEMS

There are several excellent references on the detailed analysis and design of ducting systems.[2,5,13-16] We shall not attempt to reproduce that material here but instead will confine outselves to comments on the applicability of these procedures to asbestos-handling systems.

In standard ventilation design procedure,[5] two design processes are normally available. They are the "air balance without blast gate adjustment" method and the "air balance by blast gate adjustment method." Only the first of these methods should be considered for asbestos dust system design. This is because the second system depends heavily on blast gate adjustments to the completed system to achieve proper balancing between branches. As the number of blast gates should be minimized in asbestos systems, the blast gate adjustment method should not be used.

The recommended procedure without blast gates is also called the static pressure balance method. A calculation begins by choosing the branch assumed to have the greatest resistance and proceeds toward the fan, calculating losses along this path. At each junction the static pressure for the joining branch is set equal to the calculated value at that point and the branch is then sized to give the desired flow if this static pressure is maintained. This design procedure will normally produce actual flows within 10-20% of design flows at all points in the system.

A design process used in air conditioning work called the "total pressure method" is also suitable for dust-handling design. The process is essentially one of complete analysis of a series of different geometries from fundamental considerations. Details of the method are given in the ASHRAE manual.[14] Other methods of duct design used in air conditioning practice, such as the "equal friction" and "velocity reduction" methods, are not suitable for asbestos-handling designs. Both of the recommended procedures are available as computer programs.

Ducting systems associated with air pollution control equipment have traditionally been designed to provide minimum capital cost and ease of maintenance. Increasingly, these systems are being designed to minimize total costs including energy costs. As a result, there is considerable interest in developing minimum pressure-drop ducting. The ducting associated with the equipment may be designed for minimum pressure drop by minimizing the number of loss-creating elements in the system. The following general rules summarize a considerable amount of more detailed knowledge:

1. Since losses are proportional to the square of velocities, lower velocities are to be preferred for lower losses. Since velocities must be large enough to prevent blockage, this choice is probably the most important decision to be made about any system.
2. Since the diffusion process is inherently inefficient, flow should be slowed down as infrequently as possible.
3. Smooth turns are to be preferred to sharp turns at any point where the flow must be changed in direction.
4. Efforts must be made to use all parts of the filter assembly with equal efficiency. As a result, attempts should be made in inefficient installations to distribute flow evenly over the entire filtering surface.
5. Systems designed to operate within a narrow range of operating quantities must be carefully maintained so that performance is kept up.

Design flow rates cannot be maintained if makeup air is not sufficient. All systems must provide makeup air to allow proper operation of equipment, which operates by natural convection such as chimneys and vents. This makeup air should be provided as far away as possible from dusty areas so that the flow of air in the work area be from clean regions toward the contaminating region and the hoods. This air is ordinarily fan-supplied at levels less than 10 ft above the floor. Axial flow fans are frequently used for this purpose. Failure to provide sufficient makeup air will also lower pressures in the work area with resulting problems from drafts, and difficulty in opening doors.

MONITORING AND TESTING

The subject of equipment for air and dust monitoring in ventilation systems is documented extensively in the literature, and we shall again confine ourselves to relating this material to asbestos practice.

Velocity and static pressure measurements are easily made either by hand or on a monitoring basis. The basic instruments for this purpose are the pitot static tube, the manometer or other pressure-measuring devices and various forms of anemometer and velometer.

Care must be taken to use these instruments properly, since considerable error is possible in some of them. The static pressure tapping, both on pitot static tubes and on duct and hood walls is particularly prone to error. Holes greater than 1/16 in. in diameter, holes with burrs and holes plugged with moisture or dust are common sources of error for static pressure toppings. In the use of pitot tubes it can be difficult to orient the tube directly into the stream in some locations. This is not normally a source of error since pitot tubes give accurate readings over a range of ± 10° to the flow direction. For monitoring instruments the most common problem in asbestos systems is plugging. A regular maintenance program requires that pressure tubing and static pressure toppings be regularly checked for blockage.

The velometer is the usual device used in surveying hood performance and it may also be used in many duct testing situations. Guidelines for using the velometer are available in the literature.[5] Constant use of a velometer in dusty atmospheres can alter the calibration. For this type of application, velometers can be obtained for use with a filter, and in these circumstances, the filter must always be used. As the filter loads with dust, however, the plugging becomes a source of error. Velometers should be calibrated periodically against a pitot tube to ensure their accuracy. In using velometers to measure hood performance, the measuring device must be at the hood face since velocity and static pressure gradients are steep in this region.

All duct systems should be tested at regular intervals. In many applications, exhaust ventilation, since it is not part of the production equipment, is not as well maintained as it should be. Only regular surveys can slow blockage and other poor behavior in the system. A complete survey of a system should be made when first installed. This survey should document fully the behavior of the system, and the information (velocities, static pressures, fan rpm etc.) should be kept permanently. Later comparison with these data will often be useful in finding operating problems. If velocities and flows are measured at inlet and outlet points only, and if there is no leakage, then static pressure measurements made at several points along the system can give a complete picture of the flow. It is particularly important to document static pressure drops across hoods and across the fan. This will later aid in troubleshooting, if fan problems must be separated from ducting problems.

Dust monitoring, particularly of the air downstream of the filter, is frequently recommended. While highly desirable, this requirement poses special problems for asbestos dust. Fiber counting by membrane filter procedures cannot be automated. As a result, those monitors in use are of the automatic particle count type. These are ordinarily photometers which count particles by their effect on a concentrated light beam as they pass through it. These devices are effective in recording changes in dust count and, as a result, are useful in monitoring the effectiveness of filters and can warn of such things as bag leakage. Their relationship to fiber counts, however, is in question. High-volume filters can also be mounted in plant work areas and on exhaust ducting to give gravimetric readings. Again, their relationship to fiber counts is not clear and they

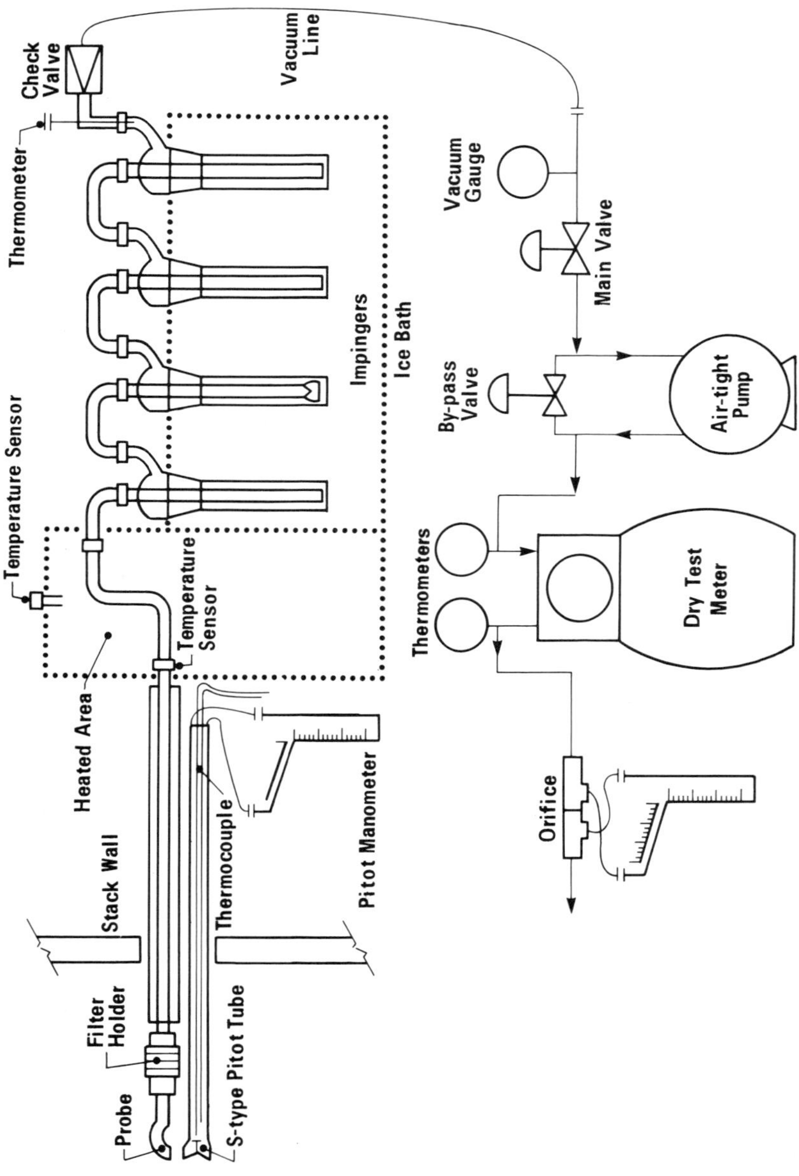

Figure 12.11. Modified stack sampling train for in-stack asbestos sampling.

serve only to record changes in overall dust levels. Recently, a laser-based device has come on the market that purports to measure fiber counts continuously. The effectiveness of this device is not known.

Because of the difficulty in monitoring dust counts, regular fiber counting by the standard MF procedure is necessary. For exhaust ducting, this can be accomplished by modifying the standard Environmental Protection Agency stack sampling train to mount on a stack membrane filter sampler.[8,17] The arrangement is shown in Figure 12.11.

REFERENCES

1. Goldfield, J., and F. E. Brandt. "Dust Control Techniques in the Asbestos Industry," *Am. Ind. Hyg. J.* 35:799-808 (1974).
2. Jorgenson, R. *Fan Engineering,* Buffalo Forge Co., Buffalo, NY (1970).
3. Bamblin, W. P. "Dust Control in the Asbestos Textile Industry," *Ann. Occup. Hyg.* 2:54-74 (1959).
4. Goldfield, J. "Fabric Filters in Asbestos Mining and Asbestos Manufacturing," Paper #74-275, 67th Annual Air Pollution Control Association Meeting, June, 1974.
5. Committee on Industrial Ventilation. *Industrial Ventilation,* 14th ed. American Conference of Fobernmental Industrial Hygienists, Lansing, MI (1976).
6. Hutcheson, J. R. M. "Environmental Control in the Asbestos Industry of Quebec," *Can. Inst. Min. Bull.* 83-89 (August 1971).
7. Milner, P. C., and J. G. Mallon. "Control of Asbestos Emissions in an Urban Environment," in *Proc. 24th Ann. Ind. Waste Conf.* (May 1977), pp. 94-116.
8. Bumbaco, M. J., B. D. Williams and J. H. Shelton. "Study of Interlaboratory Count Correlation and Fiber Distribution on Asbestos Stack Samples," Environmental Protection Service Report EPS5-AP-76-7, Department of the Environment, Ottawa (September 1976).
9. Roach, S. A. "Hygiene Standards for Asbestos," *Ann. Occup. Hyg.* 13:7-15 (1970).
10. Astleford, W. J. "Recirculation of Exhaust Air–An Overview," in *Recirculation of Exhaust Air,* 5-24, HEW publication (NIOSH) 76-186 (1976).
11. White, P. A. F., and S. E. Smith. *High Efficiency Air Filtration* (London: Butterworths, 1964).
12. Keilholtz, G. W., and G. C. Battle. "Treatment of Radioactive Airborne Wastes from Reactors," in *Air Pollution Control* Vol. 2, W. Strauss, Ed. (New York: Wiley-Interscience, 1971).
13. Alden, J. L. *Design of Industrial Exhaust Systems* (New York: Industrial Press, Inc., 1970).
14. Danielson, J. A. *Air Pollution Engineering Manual,* Environmental Protection Service, Research Triangle Park NC (May 1973).
15. McDermott, H. J. *Handbook of Ventilation for Contaminant Control,* (Ann Arbor, MI: Ann Arbor Science Publishers, Inc., 1977).

16. American Society of Heating Refrigeration and Air Conditioning Engineers. *Handbook of Fundamentals,* New York.
17. Air Pollution Control Directorate, "Standard Reference Methods for Source Testing: Measurement of Emissions of Particulates from Stationary Sources," Environmental Protection Service, Report EPS1-AP-74-1 Environmental Canada, Ottawa (February 1974).

CHAPTER 13

NONPROCESS CONTROLS

INTRODUCTION

Complete dust control by the previously mentioned processes is not always possible. Under these circumstances, respirators must be used or workers must be put in enclosures instead of the work environment. In addition, all asbestos using plants must pay careful attention to cleaning and housekeeping procedures so that dust is not allowed to accumulate. Asbestos must be carefully disposed of because of its toxicity. These methods of control are the subject of this chapter.

RESPIRATORS AND OPERATOR BOOTHS

The use of respirators is required wherever there is a possibility of dust levels above the TLV. However, it is easier to say that respirators must be used for protection against an excess amount of asbestos dust than to induce workers to wear them. It is interesting to find out why workers will not wear respirators even when they believe they should. Surveys have shown, as expected, that the most frequently cited problem is breathing resistance and physical discomfort, meaning such things as pressure on the face and head, sweat on the face, inability to chew tobacco, and tightness of harness. A few of the workers felt that the respirators interfered with their work. Workers who, for one reason or another, suffered upper respiratory impairment were especially conscious of the need to wear a respirator, but, unfortunately, these individuals had the greatest difficulty wearing them.

There are basically three types of respirators commonly used for asbestos: quarter-mask dust respirators, single-use dust respirators and powered air-purifying respirators. All have their advantages and disadvantages. For example, quarter-mask types provide very satisfactory protection, but for them to fit properly it is necessary that they be tight, and the pressure of the straps can be painful. Also, the rubber facepiece traps sweat between it and the skin, the sweat mixes with dust and the resulting irritation becomes objectionable.

The single-use respirators, relatively new in the market, partially solve the problem of facial irritation, harness tightness and breathing burden. One big advantage of this type is that they do not require maintenance as they are not used more than once. However, these respirators have not been tested extensively and the general feeling is that they provide a somewhat lower protection than the quarter mask. Aucoin[1] concludes that only one such respirator was considered acceptable by the workers who wore test units. One of the main objections was the lack of an exhalation valve and, consequently, the workers believed they were rebreathing their own exhaled breath.

The obvious advantages of powered air-purifying respirators are well known. For example, they all use an external power source rather than the workers' lungs to overcome the filter resistance and they all confront the problem of the fit of the face mask, although a variety of different approaches are used. The obvious disadvantage is the increased weight (about 4-5 lb), which the worker must carry. The NIOSH Criteria document recommends[2] the following criteria for the use of respirators:

1. For an atmosphere containing not more than 10 f/cc greater than 5 μm in length over an 8-hour average or more than 50 f/cc over any 15-minute period, a reusable or single-use, filter-type, air-purifying respirator, operating with a negative pressure during the inhalation phase of breathing, approved under the provisions of 30CFR14 (Bureau of Mines Schedule 21B) or valveless respirators providing equivalent protection shall be used.
2. For an atmosphere containing not more than 100 f/cc greater than 5 μm in length over an 8-hour average or more than 500 f/cc over any 15-minute period, a powered air-purifying, positive-pressure respirator, approved under the provisions of 30CFR14 (Bureau of Mines Schedule 21B), shall be used.
3. For an atmosphere containing more than 100 f/cc greater than 5 μm in length over an 8-hour average or over 500 f/cc for any period in excess of 15 minutes, a type C positive-pressure, supplied air respirator, approved under the provisions of 30CFR12 (Bureau of Mines Schedule 19B), shall be used.

Recently NIOSH has begun certifying equipment. A list of current certified units is available.[3]

Operator booths are also used only in the last resort. In mining practice mechanical shovels, bulldozers, haulage trucks and other mobile equipment can be airconditioned, sealed and pressurized.[4] In mills, very dusty operations such as crushing can also be isolated and the operators provided with dust-tight cabins. Air entry to these cabins would ideally be filtered by HEPA filters. Published data on the effectiveness of these enclosures is not available.

In recent years, the problem of contaminated work clothing and the danger of taking it home has come under scrutiny. One solution that has been proposed is that of disposable coveralls.[5] Certainly outer clothing used in asbestos factories should not be worn outside the factory as a general practice. Change rooms should be provided by the employer for workers regularly exposed to airborne concentrations of asbestos fibers.

Clothes lockers with two separate lockers for each worker separated by a shower room can be provided so that one locker for street clothes is in one room and a second locker for work clothes is in another room. Contaminated clothing should be placed in a covered impermeable container.

Laundering of asbestos-contaminated clothing can be done so as to prevent the release of airborne asbestos fibers in excess of the TWA. Those who give asbestos-contaminated clothing to another person or company for laundering should inform them of the presence of dust. Contaminated clothing should also be transported in sealed impermeable bags or containers and a caution label attached.

CLEANING PROCEDURES

In asbestos industries, all equipment and the plant must be kept free of dust if fiber levels are to be kept low. All necessary cleaning must be vacuum or wet cleaning, since dry sweeping and similar procedures create more rather than fewer dust problems.

One of the two forms of vacuum cleaning units used in the asbestos plants is a portable industrial vacuum cleaner using filter bags. The filtered air is returned to the working environment. The other form is a central vacuum cleaning system which consists of a central suction and filtration unit from which ducts run to those parts of the plant in which vacuum cleaning is necessary. The first form of vacuum cleaning is adequate only where asbestos is used at irregular and infrequent intervals due to the filtered air returning to the work place. On the other hand, where asbestos is regularly used in the processes, central vacuum cleaning is more efficient.

The portable units should be of the heavy-duty type. Goldfield and Brandt[6] recommend 1.5-in. or 2-in. piping and up to 25 ft of flexible hose. Static pressure in the tool should be 1 or 2 in. of mercury. These authors strongly recommend large-area fabric filters and caution against use of small paper filter bags. However, the Asbestos Research Council[7] requires that disposable bags be used and recommends two-stage filtration as in Figure 13.1. The motor for these units should be 3 hp or over driving a multiple-stage centrifugal compressor.

Central vacuuming systems are also in common use. These are often designed for a 25-ft length of hose. The systems require high-pressure drops due to the long runs of small diameter piping required. In Britain, new buildings for asbestos-using processes must have these central systems[7] as well as having all surfaces smooth for ease of cleaning.

The procedures to be followed in cleaning have been outlined in the literature.[7] For floors, daily cleaning is recommended with continuous floor cleaning in larger plants. Walls are certain to become contaminated eventually and annual cleaning is the minimum recommended. Both wet cleaning and vacuuming may be used on walls. Equipment should ordinarily be vacuum cleaned, and local exhaust should always be kept running during cleaning. Under normal circumstances, equipment should be cleaned weekly or even after every shift in dirty locations such as debagging.

The most difficult cleaning is that of overhead structures. This should be done approximately three times per year in normal situations. Scaffolding or telescoping equipment may be used, and vacuuming is the preferred method. On occasion, when cleaning creates unavoidable dust, protective clothing and respirators must be used. Under no circumstances should compressed air cleaning be used.

HANDLING AND DISPOSAL OF WASTES

Disposal of wastes and asbestos containing tailings is a particularly important problem in the mining and milling industries. Tailings disposal has been a problem for the mining industry generally; however, in the Canadian mines, the tailings have often been both close to towns and very dry.[4] The problems associated with moving tailings to dumps may be controlled by wetting the material during transfer. The problems of covering tailings with soil and creating growth on the surface are different from the normal mine tailings problem. Normal mine tailings are acidic, while asbestos tailings are highly alkaline ($pH = 9$). Attempts have been made to mix copper mine tailings with asbestos tailings to neutralize the effect on the covering soil.

In milling practice, the very short fibers collected from baghouses and other air cleaners often must be disposed of, since the fibers are too short to be of commercial value. Since this material is particularly prone to reenter the air, it must be treated with particular care.

In manufacturing plants, dust from air cleaning devices is very likely to reenter the air. Another major disposal problem in manufacturing is that of the bags in which the asbestos has been delivered. Significant amounts

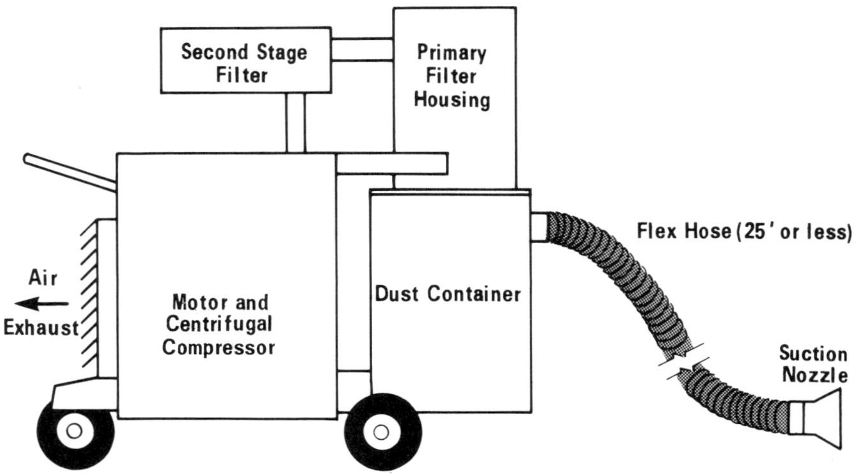

Figure 13.1. Portable vacuum cleaner with double-stage filtration.

of dust are retained in these bags and their disposal must not create additional dust. Some manufacturing processes such as asbestos-cement pipe manufacture creates wet reject asbestos and this must be treated in such a way as not to create a water pollution problem.

Demolition of asbestos-containing buildings and stripping of asbestos insulation are possible sources of air contamination. The wetting processes described previously can minimize this possibility.

Procedures for handling these various types of materials have been described by the Asbestos Research Council.[8] Their recommendations include the following.

1. Bagging outlets on hoppers should be fitted with polyethylene or other airtight bags with a tight rubber seal. 200-gauge polyethylene is recommended. Filled bags should be sealed.
2. Respirators and protective clothing should be used during all exposures to the fine dusts found in baghouses and air cleaners generally.
3. Vacuum system bags should be disposable. (However, see *Cleaning Procedures*, p.163.)
4. Material from insulation stripping or other demolition processes should be collected in plastic bags or on plastic sheeting and later sealed.
5. Larger pieces of asbestos scrap such as offcuts, breakages and rejects should be placed in containers capable of later sealing. These containers should be either sealed during transport or wetted.
6. Empty bags or sacks should be shredded under ventilation and sealed or placed in sealed bags under ventilation.
7. In Britain, crocidolite waste must be disposed of in sealed labeled bags.

Normal dumping of asbestos waste in sanitary landfill sites and other locations is permissible. However, in some cases the material should be dumped at the bottom of the site so that it can be covered to the maximum depth. As mentioned before, incineration is not ordinarily a problem, since effective incinerators have temperatures well above those at which asbestos fiber degrades.

In all places where asbestos dust is present, warning signs should be used. Also, asbestos containers should be labeled. This requirement is now compulsory in most jurisdictions.

REFERENCES

1. Aucoin, T. A. "A Successful Respirator Program," *Am. Ind. Hyg. Assoc. J.* 36:752-762 (1975).
2. Health Education and Welfare, U. S. Department of, Public Health Service. "Criteria for a Recommended Standard for Occupational Exposure to Asbestos," National Institute for Occupational Safety and Health, HSM 72-10267, Washington DC (1973).
3. Health Education and Welfare, U.S. Department of. "Cumulative Supplement NIOSH Certified Equipment," NIOSH Publication #77-195 (June 1977).
4. Hutcheson, J. R. M. "Environmental Control in the Asbestos Industry of Quebec," *Can. Inst. Min. Bull.* 64:83-89 (1971).
5. "Asbestos Workers Like Disposable Overalls," *Insulation Hyg. Prog. Rep.* 5(3) Environmental Sciences Laboratory, Mt. Sinai School of Medicine, New York (1973).
6. Goldfield, J., and F. E. Brandt. "Dust Control Techniques in the Asbestos Industry," *Am. Ind. Hyg. Assoc. J.* 36:799 (1974).
7. "The Cleaning of Premises and Plant in Accordance with the Asbestos Regulations," in *Control and Safety Guide #9,* Asbestos Research Council, London (1975).
8. "Recommended Code of Practice for the Handling and Disposal of Asbestos Waste Materials," Asbestos Research Council, London (1973).

CHAPTER 14

THE ECONOMICS OF DUST CONTROL

COSTING PRINCIPLES

To optimize a dust control system, it is necessary to do an economic analysis. This is particularly important when comparing different systems. An economic analysis should be performed when deciding on location of baghouses, types of filter material, ducting arrangements and similar questions. The method of comparison to be used will frequently be determined by company procedures but will normally be by comparing the costs of each system calculated by one of the following four methods[1];

1. present worth – the cost today of all elements of present and future costs expressed as a single sum;
2. equivalent uniform annual cash flow – the constant annual payment equivalent to all present and future costs;
3. the prospective rate of return on investment compared to a minimum attractive rate of return; or
4. the benefit/cost ratio – normally a government project cost calculation.

In each case, an assumed minimum attractive rate of return of investment must be specified as an interest rate. Properly calculated, these methods will give similar results.

While we cannot present a completely documented procedure for cost analysis, we shall attempt to provide here some basic comparative data so that approximate costs may be calculated. In this type of calculation, it is normal to expect accuracies of the order of ± 20%. Final costing of processes finally chosen should be done with much more detailed data and quotations. However, the material given should be useful for making comparisons between systems and system optimization.

All data on costs are highly sensitive to the data for which the data was gathered. We will quote dates here for all values reported. The recommended procedure for updating these costs is the Chemical Engineering

(CE) plant cost index method[2] and the U.S. Department of Labor Bureau Statistics (BLS) on wholesale price indexes. These are published on a regular basis[3] and are useful in estimating costs if costs at past dates are known. Details of the procedures used are given by Kinkley and Neveril.[4] The indexes allow calculations to be made for many different types of capital and operating costs.

CAPITAL COSTS

The capital cost of a project can be divided into:

1. basic equipment costs (baghouse, ducting, hoods, fans, etc.),
2. auxilary equipment costs (dampers, control devices, etc.),
3. tax and transportation (4-10% of a and b may be assumed if better estimates are not available),
4. installation costs (see later),
5. engineering (assumed at 10-15% of a+ b + c + d if better estimates are not available), and
6. contingency (assumed at 10% of all items above if not standardized by company policy).

As may be seen, much of the gross cost may be estimated from knowledge of the cost of the basic equipment. An example of cost data for basic baghouses is given in Table 14.1. This cost does not include bag prices, which are summarized in Table 14.2 after Kinkley and Neveril.[4] The basic air-to-cloth ratio chosen must be modified because all sections of the baghouse are not in use due to the cleaning cycle. An approximate modifier to the net cloth area is given in Table 14.3. This enables a gross cloth area to be obtained for costing of bags. Net cloth area should be used for calculation purposes in Table 14.1

Table 14.1. Costs of Baghouses as of December 1975[4]
P = Price in U.S. Dollars; A = Net cloth area in square feet.

Type	Cost[a-c]
Intermittent, Pressure-Type Mechanical Shaker (for suction case add 10%)	P = 2910 + 1.6 A
Continuous, Suction or Pressure Pulse Jet	P = 4660 + 6.6 A
Continuous, Pressure, Mechanical Shaker (for suction case add 10%)	P = 5780 + 3.0 A
Continuous Pressure Reverse Air (for suction case add 10%)	P = 22300 + 2.6 A

[a]These prices do not include bags.

[b]Add on approximately 50% if stainless steel construction specified.

[c]Add on 40-50% if insulation specified.

Table 14.2. Bag Prices Per Square Foot[4] (Data for December 1975)

Class	Type	Dacron	Orlon	Nylon	Nomex®	Glass	Polypropylene	Cotton
	Mechanical shaker, <20000 ft^2	0.35	0.60	0.70	1.10	0.45	0.60	0.40
Standard	Mechanical shaker, >20000 ft^2	0.30	0.55	0.65	1.00	0.40	0.50	0.35
	Pulse jet[a]	0.55	0.90		1.25		0.65	
	Reverse air	0.30	0.55	0.65	1.00	0.40	0.50	0.35
Custom	Mechanical shaker	0.20	0.30	0.40	0.60	0.25	0.30	0.35
	Reverse air	0.20	0.30	0.40	0.60	0.25	0.30	0.35

[a]For heavy felt, multiply price by 1.5.

Table 14.3. Approximate Guide to Estimate Gross Cloth Area[4]

Net Cloth Area (ft^2)	Gross Cloth Area (ft^2)
1 - 4,000	Multiply by 2
4,001 - 12,000	Multiply by 1.5
12,001 - 24,000	Multiply by 1.25
24,001 - 36,000	Multiply by 1.17
36,001 - 48,000	Multiply by 1.125
48,001 - 60,000	Multiply by 1.11
60,001 - 72,000	Multiply by 1.10
72,001 - 84,000	Multiply by 1.09
84,001 - 96,000	Multiply by 1.08
96,001 - 108,000	Multiply by 1.07
108,001 - 132,000	Multiply by 1.06
132,001 - 180,000	Multiply by 1.05
180,001 on up	Multiply by 1.04

Hooding and enclosures have been examined in detail by Kinkley and Neveril and their costs are approximately 20-25 cents/lb for material and 30 cents/lb of plate for labor as of December 1975. Ducting is also treated in this work and costs for 1/8 in. thick carbon steel straight duct are given as:

$$P = -1.25 + 0.93D \tag{1}$$

where P is the price in U.S. dollars per linear foot and D is the duct diameter in inches. Again, the equation for December 1975 and is valid for duct diameters over 12 in. The costs of elbows, tees, dampers, etc. are also given by Kinkley and Neveril.[4] Fan costs for backward-bladed fans suitable for use on the clean side of a dust collection system and producing 5 - 6 in. of water pressure are given in Table 14.4 as of January 1977. Motor costs run at approximately 20-30 % of fan costs for normal installations. Other capital expenditures such as dust disposal equipment, cyclones, etc., are dealt with in the literature.[4,5]

The installation costs of control systems are quoted over wide ranges depending on local conditions.[4,5] For bag filters, these can range from 40-120 % of basic capital costs. A more normal figure should be about 75 or 80% and includes installation of motors, fans, foundations, etc. High installation costs can be expected for retrofitting of equipment. Factors to be considered in estimating installation costs include:

1. location and ease of approach for larger pieces of equipment;
2. available space for equipment and the possibility of crowding;
3. amount of instrumentation desired;

4. degree of testing and performance checking to be done on the installation;
5. degree to which the installation is a nonstandard design;
6. amount of assembly required onsite; and
7. cost of local wages.

Table 14.4. Approximate Costs of Backward Bladed Fans Developing 5-6 in. Pressure Drop (Data as of January 1977)

Flow (cfm)	Cost (US $)
5,000	900
10,000	1,400
20,000	3,000
50,000	7,000
100,000	13,000

OPERATING COSTS

The major operating costs for most dust control systems are the electrical cost for fan and filter operation and the maintenance costs. The electrical cost for one year of operation of a fabric filter system may be estimated from[6]:

$$C = 1.173 \times 10^{-4} \frac{SPHK}{E} \qquad (2)$$

where C is the annual electrical cost in U.S. $
S is the design capacity of the system in cfm (actual)
P is the pressure drop in in. of water
H is the estimated yearly hours of operation
K is the cost of electricity in dollars/kWh
E is the fan efficiency expressed as a percentage

Electrical costs may also be estimated by determining the expected operating point of the fan and calculating basic horsepower requirements.

Maintenance costs have been estimated by EPA as approximately 5¢/cfm/yr in 1973 with a range of 3-10¢. Kinkley and Neveril[4] have estimated maintenance costs for fabric filter systems at approximately 2% of equipment purchase price with a range from 1-6%. They further suggest that normal bag life is around 1.5 years. Goldfield[7] suggests that a bag life of 6-7 years is possible. The expected life of the fabric filter system has been estimated at 20 years with a range from 5-40 years.

Makeup air heating can be a major operating expense. This is because of the necessity of heating the air in most climates. The *Industrial Ventilation*[8] manual gives equations for estimating this cost. The estimate is based on an efficient makeup air system and will give low results if the air enters by infiltration. The cost per hour of running is estimated as:

$$\text{Hourly cost} = \frac{0.001\,QNc}{q} \tag{3}$$

where Q is the air flow in cfm
N is the required heat in Btu/hr/1000 cfm
c is the cost of fuel in dollars per unit
q is the available heat per unit of fuel

The value of N depends on the outside air temperature, and values are given in Table 14.5.

Table 14.5. Heat Required to Provide 70°F Air

Average Outside Air Temperature (°F)	N, Required Heat in Btu/hr/1,000 cfm to Provide 70°F Air
0	75,500
5	70,000
10	65,000
15	59,500
20	54,000
25	48,500
30	43,000
35	38,000
40	32,500
45	27,000
50	21,500
55	16,000
60	11,000
65	5,500

The amount of heat available from various types of fuel is needed to evaluate q. Estimates of these values are given in Table 14.6.

Estimates of average yearly cost are also given in the *Industrial Ventilation*[8] manual. The procedure uses the degree-day method common to air conditioning practice.

Table 14.6. Estimated Heat Available from Various Fuels

Fuel	Btu/Unit	Efficiency (%)	Available BTU Per Unit (q)
Coal	12,000 Btu/lb	50	6,000
Oil	142,000 Btu/U.S. gal	75	106,500
Gas			
Heat exchanger	1000 Btu/ft^3	80	800
Direct fired		90	900

GENERAL COMMENTS ON COSTS

Costing projects in the general way which we have attempted here can only be an approximate process. Many more variables than those discussed will influence the costs. Goldfield[7] lists the following factors as significant influences on collector costs:

1. size of system
2. duct construction
3. use of cyclones ahead of fabric filters
4. intermittent or continuous filtration
5. type of cloth in the filter
6. insulation of the filter
7. material handling equipment for dust discharges
8. controls

In preparing gross estimates of dust handling costs, the capital cost of collecting and cleaning one cfm of air is often requested. This has been variously reported as 2-6 \$/cfm[7,8] for the years 1973 and 1974, respectively. This large range is, of course, a result of the large number of approximations necessary in arriving at such a global figure.

REFERENCES

1. Grant, E.L., and W.G. Ireson. *Principles of Engineering Economy,* 5th ed. (New York, The Ronald Press, 1970).
2. Arnold, T.H., and C.H. Chilton. "New Index Slows Plant Cost Trends," *Chem. Eng.* 143-149. February 18, 1963).
3. "Economic Indicators," *Chem. Eng.*
4. Kinkley, M.L., and R.B. Neveril. "Capital and Operating Costs of Selected Air Pollution Control Systems, U.S. Environmental Protection Agency report EPA-450/3-76-014 National Technical Information Service, Springfield, VA (May 1976).

5. "Control Techniques for Asbestos Air Pollutants," U.S. Environmental Protection Agency, AP117 Research Triangle Park, NC (1973).
6. Edmisten, N.G. and F.L. Bunyard. "A Systematic Procedure for Determining the Cost of Controlling Particulate Emissions from Industrial Sources, *J. Air Poll. Control Assoc.* 20(7): 446-452 (1970).
7. Goldfield, J. "Fabric Filters in Asbestos Mining and Asbestos Manufacturing," Paper #74-275, 67th Annual Air Pollution Control Association Meeting, (June, 1974).
8. Committee on Industrial Ventilation. *Industrial Ventilation,* 14th ed., American Conference of Government Industrial Hygienists, Lansing, MI (1976).

APPENDIX I

MAJOR PROCESSORS OF ASBESTOS IN THE U.S. IN 1968 (REFERENCE 5, CHAPTER 2)

REPROCESSING PLANTS PRODUCING ASBESTOS FRICTION MATERIALS

	Location
CALIFORNIA	
H. Krasne Manufacturing Co.	Los Angeles
Silver Line Brake Lining Corp.	Los Angeles
Lasco Brake Products Corp., Ltd.	Oakland
CONNECTICUT	
H. K. Porter Company, Inc.	Middletown
Raybestos-Manhattan, Inc.	Stratford
ILLINOIS	
Gatke Corp.	Chicago
Grizzly Brake Division of Mar Pro Inc.	Chicago
The L. S. Miley Co.	Chicago
Johns-Manville Corp.	Waukegan
INDIANA	
Raybestos-Manhattan, Inc.	Crawfordsville
H. K. Porter Co., Inc.	Huntington
World Bestos Co. (sub. of Firestone Tire & Rubber Co.)	New Castle
KENTUCKY	
H. K. Porter Co., Inc.	Richmond
MASSACHUSETTS	
Auto Friction	Lawrence
MICHIGAN	
American Brakeblock - Division Ambes Corp.	Birmingham
American Brake Shoe Co.	Troy

	Location
NEW JERSEY	
Johns-Manville Corp.	Manville
Reddaway Manufacturing Co., Inc.	Newark
Raybestos-Manhattan, Inc.	Passaic
H. K. Porter Co., Inc.	Trenton
NEW YORK	
Bendix Corp.	Troy
NORTH CAROLINA	
Southern Friction Materials Co.	Charlotte
OHIO	
American Brake Shoe	Cleveland
General Motors Corp.	Dayton
Maremont Corp.	Paulding
PENNSYLVANIA	
Raybestos-Manhattan, Inc.	Manheim
H. K. Porter Co., Inc.	Pittsburgh
TEXAS	
Standco Brake Lining Co.	Houston
TENNESSEE	
Bendix Corp.	Cleveland
VIRGINIA	
American Brake Shoe	Winchester

REPROCESSING PLANTS PRODUCING ASBESTOS CEMENT PRODUCTS

ALABAMA	
U.S. Cast Iron Pipe Co.	Anniston
GAF Corp.	Mobile
Cement Asbestos Products Co.	Woodward
CALIFORNIA	
Southern Pipe & Casting Co.	Azuza
Johns-Manville Corp.	Long Beach
Johns-Manville Corp.	Pittsburgh
Certain-Teed Products Corp.	Riverside
Certain-Teed Products Corp.	Santa Clara
Johns-Manville Corp.	Stockton
CONNECTICUT	
Tile Roofing Co.	Stratford
GEORGIA	
Uniroyal, Inc.	Hogansville
GAF Corp.	Port Wentworth

	Location
ILLINOIS	
Acme Asbestos Covering & Flooring Co.	Chicago
Asbestos & Magnesia Materials Co.	Chicago
Flintkote Co.	Chicago
Western Slate Co.	Elmhurst
Fel-Pro, Inc.	Skokie
Johns-Manville Corp.	Waukegan
INDIANA	
U.S. Gypsum Co.	East Chicago
LOUISIANA	
Johns-Manville Corp.	Marraro
National Gypsum Co.	New Orleans
MASSACHUSETTS	
Johns-Manville Corp.	No. Billerica
MICHIGAN	
American Asbestos Products Co.	Detroit
MINNESOTA	
Mac Arthur Co.	St. Paul
MISSOURI	
Certain-Teed Products Corp.	St. Louis
GAF Corp.	St. Louis
National Gypsum Co.	St. Louis
NEW HAMPSHIRE	
Johns-Manville Corp.	Nashua
NEW JERSEY	
Johns-Manville Corp.	Manville
National Gypsum Co.	Millington
Philip Carey Manufacturing Co.	Perth Amboy
GAF Corp.	So. Bound Brook
U.S. Plywood-Champion Papers, Inc.	South River
NEW YORK	
Asbeka Fabrications Corporation	Brooklyn
National Gypsum Company, Inc.	Buffalo
Flintkote Co.	White Plains
NORTH CAROLINA	
H. K. Porter Co., Inc.	Charlotte
Johns-Manville Corp.	Marshville
OHIO	
Philip Carey Manufacturing Co.	Cincinnati
Seagrave	Columbus
Flintkote Co.	Ravenna
PENNSYLVANIA	
Certain-Teed Products Corp.	Ambler
Nicolet Industries, Inc.	Ambler
Supradur Manufacturing Co. (sub. of Seagrave)	Windgap

	Location
TEXAS	
Johns-Manville Corp.	Dennison
Certain-Teed Products Corp.	Hillsboro
Philip Carey Manufacturing Co.	Houston
WISCONSIN	
Wisconsin Gasket & Manufacturing Co.	Milwaukee

REPROCESSING PLANTS USING ASBESTOS IN FLOOR TILE

CALIFORNIA	
GAF Corporation	Long Beach
Armstrong Cork Co.	South Gate
Flintkote Co.	Vernon
ILLINOIS	
Flintkote Co.	Chicago
GAF Corp.	Joliet
Armstrong Cork Co.	Kankakee
Johns-Manville Corp.	Waukegan
LOUISIANA	
Johns-Manville Corp.	Marraro
Flintkote Co.	New Orleans
MASSACHUSETTS	
Flintkote Co.	Watertown
MISSISSIPPI	
Armstrong Cork Co.	Jackson
NEW JERSEY	
Congoleum-Nairn, Inc.	Kearny
Johns-Manville Corp.	Manville
American Builtrote Rubber Co., Inc.	Trenton
NEW YORK	
GAF Corp.	Vails Gate
OHIO	
Johns-Manville Corp.	Chillicothe
PENNSYLVANIA	
Armstrong Cork Co.	Lancaster
TEXAS	
GAF Corp.	Houston

OTHER ASBESTOS REPROCESSING PLANTS*

	Location	Code
ALABAMA		
GAF Corp.	Mobile	B

*CODE: A-Textiles, B-Paper, C-Miscellaneous

	Location	Code
CALIFORNIA		
Hill Brothers Chemical Co.	City of Industry	
Fibreboard Paper Products Corp.	Emeryville	C
Johns-Manville Corp.	Pittsburgh	BC
Fibreboard Paper Products Corp.	Redwood City	B
Scott Labs, Inc.	Richmond	C
Sacomo Manufacturing Co.	San Francisco	ABC
George Short Co.	San Francisco	C
COLORADO		
The Verticel Co.	Englewood	
CONNECTICUT		
Brosites Industries, Inc.	Greenwich	
Standard Washer & Material, Inc.	Manchester	C
Auburn Manufacturing Co.	Middletown	C
DELAWARE		
Haveg Industries, Inc.	Wilmington	
GEORGIA		
Terri-Cord Mills	Roberta	
ILLINOIS		
Unarco Industries, Inc.	Bloomington	C
Accurate Felt & Gasket Manufacturing Co.	Chicago	C
Acme Asbestos Covering & Flooring Co.	Chicago	B
Asbestos & Magnesia Materials Co.	Chicago	AB
Asbestos Textile Co.	Chicago	A
Chambers Gasket & Manufacturing Co.	Chicago	C
Colonial Kolonite Co.	Chicago	
A. Daigger & Co.	Chicago	C
Filpaco Industries, Inc.	Chicago	C
Geraghty Gasket & Manufacturing Co.	Chicago	C
L. E. Harnisch & Co.	Chicago	BC
John Herman Manufacturing Corp.	Chicago	A
Industrial Hermetic Materials, Inc.	Chicago	C
Kopel Filter Paper Co.	Chicago	C
Sall Mountain Co.	Chicago	C
Singer Safety Products, Inc.	Chicago	A
Grant Wilson, Inc.	Chicago	AB
United Gasket Corp.	Cicero	C
Industrial Gloves Co.	Danville	A
GAF Corp.	Joliet	B
F. D. Farnam Co.	Lyons	ABC
Luse-Stevenson Co.	Melrose Park	C
Blackhawk Gasket Corp.	Rockford	C
Excelsior Leather Washer Manufacturing Co.	Rockford	C
Nicolet Industries, Inc.	Union	
Johns-Manville Corp.	Waukegan	BC
LOUISIANA		
Johns-Manville Corp.	New Orleans	B

	Location	Code
MARYLAND		
Congoleum-Nairn, Inc.	Cedarhurst	B
MASSACHUSETTS		
Armstrong Cork Co.	Braintree	C
Pepperell Braiding Co.	East Pepperell	A
Barwood Manufacturing Corp.	Everett	C
Asbestos Textile Company, Inc.	No. Brookfield	A
MICHIGAN		
Detroit Gasket & Manufacturing Co.	Detroit	B
Great Lakes Filter Media Co.	Detroit	C
Walker Manufacturing Co.	Jackson	
MISSOURI		
GAF Corp.	Kansas City	B
Standard Asbestos Manufacturing & Insulating Co.	Kansas City	BC
General Gasket Corp.	St. Louis	C
Tallman-McCluskey Fabrics Co. (sub. of H. K. Porter Co. Inc.)	St. Louis	A
NEW HAMPSHIRE		
American Asbestos Textile Corp.	Meredith	A
Johns-Manville Corp.	Tilton	BC
NEW JERSEY		
Howard Industries, Inc.	Berkley Hgts.	
Asbesto Corp.	Bloomfield	A
Kavon Filter Products Co.	Cranford	C
Janos Asbestos Co.	East Rutherford	ABC
GAF Corp.	Gloucester City	B
Columbia Filter Co.	Hawthorne	C
Cellulo Co.	Hoboken	C
Imperial Products Co.	Hoboken	
Ladden Asbestos Corp. of New Jersey	Irvington	AB
Smyth Rubber & Packing Co.	Jersey City	C
Smith & Kanzler Corp.	Linden	BC
Johns-Manville Corp.	Manville	BC
Flaherty-Kennedy Filter Fabrics	Maplewood	C
Asbestos Products Manufacturing Corp.	Newark	C
Asbestospray Corp.	Newark	C
Johns-Manville Corp.	New Brunswick	C
Electrical Insulation Sales Co.	North Bergen	C
Brassbestos Manufacturing Corp.	Patterson	
La Favorite Rubber Manufacturing Co.	Patterson	C
J. T. Baker Chemical Co.	Phillipsburg	
Minerals & Insulation Co., Inc.	Rochelle Park	C
Baldwin-Ehret-Hill, Inc.	Trenton	C
NEW YORK		
Atlantic Asbestos Corp.	Bronx	BC
Able-Val Canvas & Rope Manufacturing Co.	Brooklyn	A
Acme Canvas & Rope Co.	Brooklyn	A
Eureka Packing Co.	Brooklyn	A

	Location	Code
Smith Chemical & Color Co.	Brooklyn	
U.S. Indestructible Gasket Co.	Brooklyn	AC
Armstrong Cork Co.	Fulton	B
National Filter Corp.	Mt. Vernon	C
Asbestos Ltd.	New York City	
Atlantic Asbestos Corp.	New York City	
Scientific Filter & Machinery Co.	New York City	
Whittaker, Clark & Daniels, Inc.	New York City	
Pam Narrow Fabrics Corp.	Oceanside	A
Garlock, Inc.	Palmyra	AC
Gaddis Engineering Co.	Port Washington	BC
NORTH CAROLINA		
Carolina Asbestos Co.	Davidson	A
Tar Heel Mica Co.	Plumtree	
OHIO		
G. P. Hall Co.	Akron	
Johns-Manville Corp.	Avery	B
The Blemker Co.	Cincinnati	
Cincinnati Gasket Packing & Manufacturing Co.	Cincinnati	C
Russel Gasket Co.	Cincinnati	C
Zimmerman Packing Co.	Cincinnati	C
Asbestos Products Co.	Cleveland	
Bodwell-Lemmon Co.	Cleveland	
Foseco Inc.	Cleveland	C
Russell Gasket Co.	Cleveland	C
Nicolet Industries, Inc.	Hamilton	
The Cellulo Co.	Sandusky	C
Asbeka Fabricators Corp.	Willoughby	C
PENNSYLVANIA		
Keasbey & Mattison Co.	Ambler	AC
Nicolet Industries, Inc.	Ambler	C
GAF Corp.	Erie	AC
Debco Products	Herminie	B
Raybestos-Manhattan, Inc.	Manheim	AC
American Asbestos Textile Corp.	Norristown	A
Nicolet Industries, Inc.	Norristown	
Atlas Asbestos Co.	North Wales	A
Greene Tweed & Co.	North Wales	C
Aljay Manufacturing Co.	Philadelphia	A
Anchor Packing Co.	Philadelphia	C
Austen Hill Manufacturing Co.	Philadelphia	
Burnwick Asbestos Co.	Philadelphia	
Collins Packing Co.	Philadelphia	C
Delaware Asbestos & Rubber Co.	Philadelphia	
Manufactured Rubber Products Co.	Philadelphia	
Mercer Rubber Co., Inc.	Philadelphia	A
Philadelphia Asbestos Corp.	Philadelphia	AC
George A. Rowley & Co., Inc.	Philadelphia	

	Location	Code
Charles A. Wagner Co., Inc.	Philadelphia	
Armstrong Cork Co.	Pittsburgh	C
Pittsburgh Corning Corp.	Pittsburgh	
Westinghouse Electric Corp.	Pittsburgh	C
Refractory and Insulation Corp.	Port Kennedy	C
Quaker Safety Products & Manufacturing Co.	Quakerstown	
Carlisle Corp.	Ridgway	
SOUTH CAROLINA		
H. K. Porter Co., Inc.	Bennettsville	A
Raybestos-Manhatten, Inc.	North Charleston	AC
TEXAS		
GAF Corp.	Dallas	B
Johns-Manville Corp.	Fort Worth	B
Philip Carey Manufacturing Co.	Houston	A
Standco Asbestos Textile Co.	Houston	AC
VIRGINIA		
Capitol Asbestos Fabricators Corp.	Alexandria	
Forcee Manufacturing Corp.	Tappahannock	
WISCONSIN		
Ametek/Plymouth Plastics Division	Sheboygan	

APPENDIX II

MAJOR PROCESSORS OF ASBESTOS IN CANADA 1976 (REFERENCE 6, CHAPTER 2)

Asbestos Product Manufacturers of Canada[a]

Companies	Locations	Principal Products
British Columbia		
Dura Pipe Co., Ltd.	Surrey	Asbestos cement
Home Insulation Ltd.	Surrey	Pipe materials
Synkoloid Asbestos Products Ltd.	Surrey	Asbestos cement sheet
Turners Building Products	Mission	Asbestos cement sheet
Ontario		
Abex Industries Ltd.	Lindsay	Friction materials
Able Gasket and Materials Ltd.	Weston	Custom fabrication, gaskets
Canadian Durabla Ltd.	Belleville	Gasket sheets
Canadian Johns-Manville Co. Ltd.	Port Union	Asbestos cement products
Certified Automotive Products	Rexdale	
Columbia Acoustics and Fireproofing Co.	Mimico	Fireproofing materials insulations
Flintkote Co. of Canada Ltd.	Toronto	Vinyl asbestos floor covering
Garlock of Canada	Toronto	Packings and gasket sheets
Hill Machine and Asbestos Products	Downsview	Construction materials
Holmes Insulation Ltd.	Sarnia	Insulation materials
Insul-Coustics Ltd.	Blossom Park	Insulation materials
Mintex Federal	Rexdale	Friction materials
Raybestos-Manhattan (Canada) Ltd.	Peterborough	Friction materials and textiles
Scott Laboratories Ltd.	Pickering	Filters
Universal Insulations Company Ltd.	Aurora	Electrical insulating materials

Asbestos Product Manufacturers of Canada (Cont.)

Companies	Locations	Principal Products
Quebec		
Albany Felt	Cowansville	Dryer felts
Albion Asbestos Packing Ltd.	Montreal	Packings and asbestos products
Amtico Flooring Canada	Montreal	Vinyl asbestos floor coverings
American Biltrite Rubber Co. (Canada) Ltd.		
Anchor Packing Co. Ltd.	Sherbrooke	Floor coverings
Armstrong Cork (Canada) Ltd.	Montreal	Packings and gaskets
Asbestos Crop. Ltd.	Montreal	Floor coverings
Asbestos Covering Co.	Montreal	Friction materials, textiles
Asten Hill Ltd.	Montreal	Asbestos goods
Atlas Asbestos Co., Division of Bell Asbestos Mines Ltd.	Valleyfield	Dryer felts and yarn
Ayers	Montreal	Asbestos cement products
B & F Enterprises Inc.	Lachute	Dryer felts
Canadian Asbestos	Longueuil	
Canadian Johns-Manville Co. Ltd.	Ville St. Laurent	Asbestos products
Cascade Paper	Asbestos	Friction materials, textiles
Domco Industries Ltd.	Kingsley Falls	Paper and felt
Joseph Robb Co. Ltd.	Montreal	Vinyl asbestos floor coverings
Les Industries Vadco Inc.	Montreal	Packings
Mevco Products	Marieville	Pipe covering, insulations
Scapa Dryers	Ville St. Pierre	Gaskets
	Joliette	Dryer felts

[a] National Health and Welfare, Canada, Department of. Report of the Asbestosis Working Group, Subcommittee on Environmental Health (February 1976).

Conversion Factors–Foot-Pound System to Metric

To Convert From	To	Multiply By
Length		
ft	meters	0.3048
in.	meters	2.54×10^{-2}
Velocity		
fps	mps	0.3048
fpm	mps	5.08×10^{-3}
Volume flow		
cfs	$m^3 s^{-1}$	28.32×10^{-3}
cfm	$m^3 s^{-1}$	4.72×10^{-4}
Mass		
lb	kg	0.4536
gr	kg	64.8×10^{-6}
Density		
lb/ft^3	kg/m^3	16.02
gr/ft^3	g/m^3	2.29
Pressure		
$lb_f/in.^2$	$Pa(N/m^2)$	6.895×10^3
in water gauge (wg)	$Pa(N/m^2)$	249.1
Energy		
kWh	J	3.60×10^6
Power		
hp	W	745.7

Conversion Factors–Metric to Foot-Pound System

To Convert From	To	Multiply By
Length		
meters	ft	3.281
meters	in.	39.370
Velocity		
mps	fps	3.2808
mps	fpm	196.85
Volume flow		
$m^3 s^{-1}$	cfs	35.3
$m^3 s^{-1}$	cfm	2118
Mass		
kg	lb	2.2046
kg	gr	15432
Density		
kg/m^3	lb/ft^3	0.06243
g/m^3	g/ft^3	0.437
Pressure		
$Pa(N/m^2)$	$lb_f/in.^2$	1.45×10^{-4}
$Pa(N/m^2)$	in wg	4.014×10^{-3}
Energy		
J	kWh	2.78×10^{-7}
Power		
W	hp	1.341×10^{-3}

INDEX